S.J. Hedges graduated from Indiana University, Bloomington, in fine arts in 1978, where she was influenced by Mary Ellen Solt, the concrete poet. Ms. Hedges was also influenced by the poetry of Rumi and Pablo Neruda.

CATACLYSMIC EVENTS

LOVE POEMS FOR TIME TRAVELERS

S.J. HEDGES

AUSTIN MACAULEY PUBLISHERS™

Copyright © S.J. Hedges (2021)

All rights reserved. No part of this publication may be reproduced, distributed, or transmitted in any form or by any means, including photocopying, recording, or other electronic or mechanical methods, without the prior written permission of the publisher, except in the case of brief quotations embodied in critical reviews and certain other non-commercial uses permitted by copyright law. For permission requests, write to the publisher.

Any person who commits any unauthorized act in relation to this publication may be liable to criminal prosecution and civil claims for damages.

Ordering Information
Quantity sales: Special discounts are available on quantity purchases by corporations, associations, and others. For details, contact the publisher at the address below.

Publisher's Cataloging-in-Publication data
Hedges, S.J.
Cataclysmic Events

ISBN 9781649792327 (Paperback)
ISBN 9781649792334 (Hardback)
ISBN 9781649792341 (ePub e-book)

Library of Congress Control Number: 2021907625

www.austinmacauley.com/us

First Published (2021)
Austin Macauley Publishers LLC
40 Wall Street, 33rd Floor, Suite 3302
New York, NY 10005
USA

mail-usa@austinmacauley.com
+1 (646) 5125767

Dedicated to my beloved Jeffrey, who makes all things possible.

I am thankful to, and for, all the generous individuals who have helped me along my way. I am ever grateful for the support, advice, and encouragement I received in this life from teachers, friends, and especially from my beloved.

Table of Contents

Shooting Star	8
Dark of Night	10
The Dreamcatcher	12
Dream of a Distant Past	14
Fire, Earth, Water, Wind	16
Dreaming on a Warm Earth	18
Borne of the Stars	20
Stillness Borne of Motion	22
Warmed by the Fire	24
Light of the Moon	26
Blazing Blue Heavens	28
Tsunami	30
Full Moon in the Silver Night Fog	32
A Comet Appears During a Blue Moon	34
Eclipse at the Gorge	36
Solar Storm	38
Deep Cold Fog	40
The Waves	42
Solar Dream	44
Safety in the Storm	46
Remembering	48
A New Moon	50
Point/Counterpoint	52
Deep Floodwaters	56

Singularity	58
Lightning Bolt	60
Earth Quake	62
Lightning Storm	64
Golden Sunshine	66
Desert Wind	68
Eclipse of the Red Sun	70
Ice Storm	72
Magma	74
Black Hole	76
Deep Freeze	78
My Father Is the Wind	80
Urban Mechanization	82
Harsh Light	84
Dream of Happiness	86
As a Comet Appears	88
Searching for Crystals	90
To Great Spirit	92
Moment to Moment	94
Mid-Flight Standstill	96
Crisp Morning Air	98
Solar Expansion	100
Cosmic Traveler	102
Companion Planets	104
Lunar Eclipse	106

Shooting Star

From the Dwelling of my Spirit Guides
A new Star appears in the Night Sky
Heralded by Sun Spots and Floods,
Lunar eclipses and Earth changes,
Yet, I remain in the Sun.

I commune with Spirit and drink in the Sangre de Cristos
where double rainbows line this sage-covered Valley.
Tumbleweeds hide me in a private Oasis
where I know a secret cave.

Moving water fills the air
Appearing from deep within the Earth
To remind me of my communion
within the Heavens.

Dark of Night

In the dark of this night,
I am remembering again.

Candles flicker in this room where I write.
In my private sanctuary –
the one I let you enter,
I watch my dogs circle for a place to lie,
finding a familiar place,
with the warm aroma of the known.

I am remembering back when,
My heart in my throat again,
I vibrate out of my skin till when
You come to me.

The Dreamcatcher

You handled me the symbol,
then I knew, it was a message. A calling
card from the past. My cue to remember. Yes, I
have seen you before. I am magnetically drawn to your
kindred soul; the Light of God is within you.   I see
your strong face. In my dreams, I will travel to you.   I
will come on a silent stallion, to bring you sacred herbs
and lay honey at the Altar in your dreams. You are a
Warrior, I agree. You love me within energy fields.
You are not here, but ever present, always
in my heart and dreams, catching
me in your web.

Dream of a Distant Past

I know you.
Look into my eyes and remember me.
I am a woman you have known.
Those piercing warrior eyes, they are so clear.
I see your knowing, and I know you.
I have waited long for you to wrap me once again
in the magenta vibration
of the Maya and Egyptians,
the Anasazi and Atlanteans.
Long ago in Colorado, in the mountains and rivers,
we swam together, we loved together.
You gave me a stallion covered in furs
with bags made of skins for me to collect herbs.
In a lush primal glen,
in a forest, high in the mountains, we sat beside a fire.
Birds sang, dew wet our skin,
We drank tea from flowers.
In the marshes and fields of sage,
you fed me and bathed me in your light.
A velvet cape of violet energy wrapped me safely
while you held me at your fire.
You honored me and protected me,
You made a dreamcatcher.

Fire, Earth, Water, Wind

The green forest gives us our life as
Mother Earth provides all
Sacred Waterfalls renew our spirit and
We touch God in the Blue Sky.

Beyond shape and form, we come together in Light
You support me with Energy. We move together

Beyond shape and form, I rise above,
Transcending the elements
This Blue Stratosphere is my domain

Dreaming on a Warm Earth

I will visualize a healing,
a special place to rest,
to lie next to you
and dream of a future love,
to dream of a future, to dream.

I will dream to be with you
in the warm azure light
And walk among hibiscus blooms,
On a craggy mountainside where
we will eat papaya and watch whales.

On this warm earth I am dreaming,
we will observe the Moon and Stars,
the Ocean and Earth,
and move together with the elements
with which we are One.

Borne of the Stars

In the midnight skies,
between the Hunter and the Queen,
A comet appears to guide me.
In an intentional dream, we tryst in a cosmic safehouse.
There, I felt the connection,
and tonight I will return.

I travel to you, a Volcanic Being,
Ethereal energy whispers from your crown,
Surrounding and protecting you.

Your body granite,
Forged of primal fire and
Quickened by the elements,
Is inhabited by Spirit, solid as crystal
More healing, more vibrant,
My Osiris, You are borne of the Stars.

Stillness Borne of Motion

Dear Magician, how did you make time stand still?

One kiss on the hand, a quiet kiss,
As strong as the orbit of a planet
Existing between the flashes of our electrons.

Gaia Revolts:
    The Sun surges,
        The Moon responds –

But in this time, you took my hand
And held me in the moment…
No storms, no heartbeat, no breathing
Only warm steady Light and Stillness.

Warmed by the Fire

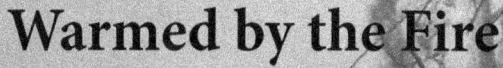

In a quiet moment, I dreamt of the night
I will sit on red silk and serve you dinner.

At a wooden table, we will sit together,
Beside the fire, with flowers on the Altar
and Sage in the Air.
We will pray for all life.

My Sacred Pottery filled with ancient grains,
potatoes, and fruits.

With Rose Petals and Sacred Herbs,
I will fill your senses.
I will warm you by my Fire.

Light of the Moon

When you left as you held me
there was a Light Gift.
As you held me, as you held me –
I began to feel your strength as a wave of Light force,
Coming through your Life source

Radiating warm, then encompassing me fully.
Your light filled me with quiet warmth
and stayed when you had gone
Keeping me gently.
In return, I send you the light of the Moon.

Blazing Blue Heavens

Before me, I see you
A dazzling comet
Blazing across blue heavens
Obscuring all other stars in the sky.
A volcano erupting all fire, no smoke,
Releasing Elements to the Heavens.
As man and God swim together in one energy body.

You are primal, enigmatic,
fascinating, mysterious, hidden, dark,
Light, loving, open,
Beautiful, stunning, Intellectual.
You are that which is Great and Originating.
A dazzling comet blazing
across blue Heavens.
You are Yang.

Tsunami

Deep in my Heart
You Happen to Me.
The deep vibration of you
Magenta and Violet

I close my eyes and experience
The tsunami current of you
Rolling thunder, rushing
waves through me.
Time after Time
Life after Life
I vibrate to you.

Full Moon
in the Silver Night Fog

If you could hear the song of my Heart,
You would believe in my Spirit.

If you could receive my message
through this night fog,
it would warm and comfort you.

I would wrap you in golden sunlight,
Even as the full moon now
wraps me in silver knowingness.

From this Mountain, Full of Energy,
I will sing to you the Song of the Crane,
And I will dance the dance
of the Wild Bird in Love.

A Comet Appears During a Blue Moon

The first time I saw you
Electricity burned through me.

My Heart opened to your Light
Then each time,
        every time,
                Same as the first time,
                      I saw you.

You, an Angel appearing through the fog,
telling me of a comet.
      An Omen.
             In the starry canopy of Night,
                  appearing between
                        the Queen and the Hunter.

On this Mountain,
Crisp Air and Blue Night
      Come down around my shoulders
           like a cool silken shawl, and envelop me
                While I receive your message of Light.

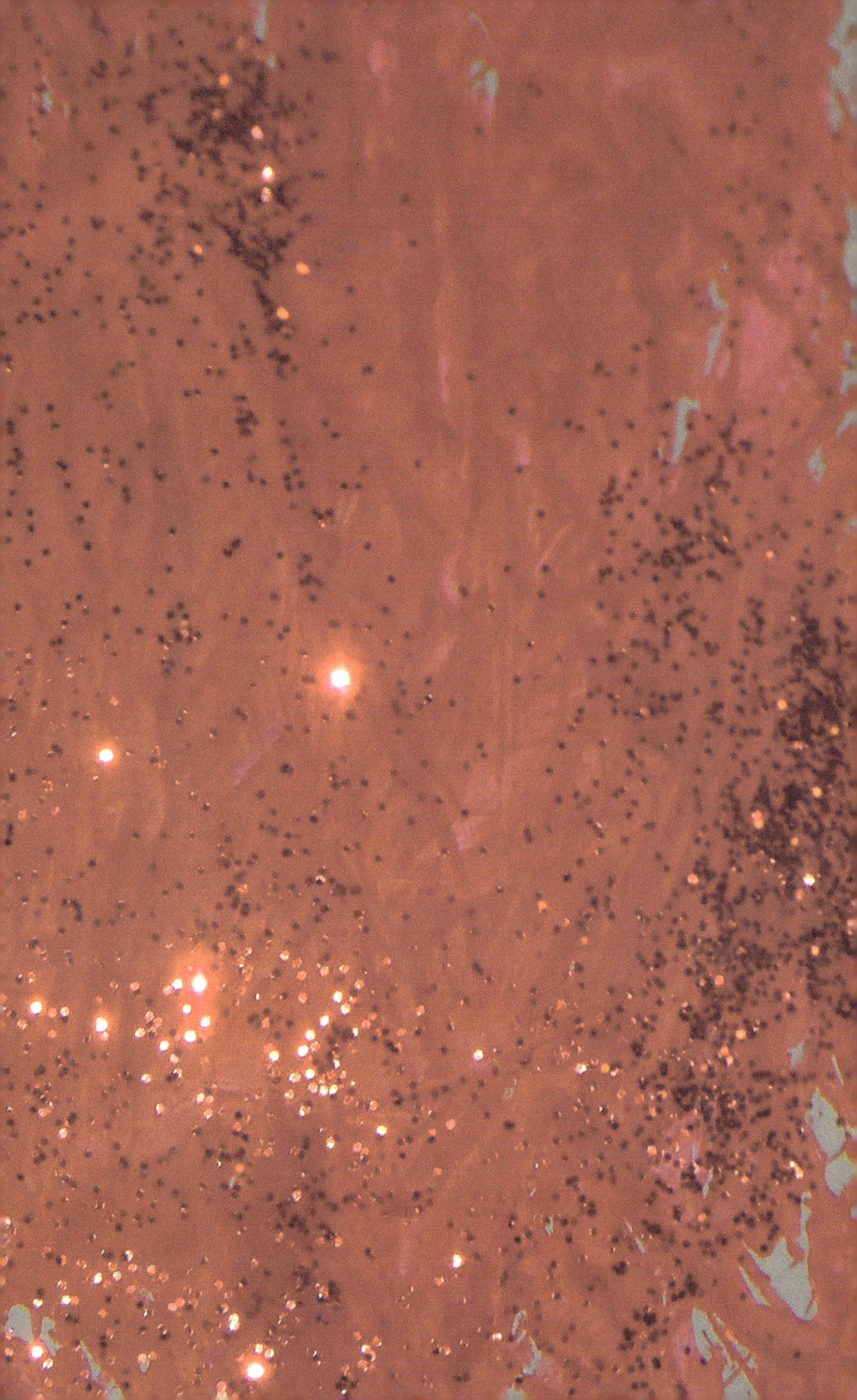

Eclipse at the Gorge

We arrived in darkness.
At a deep gorge
Cut by an ancient river,
     winding its way along the spine of the Earth.

The unconscious advice to my friend,
     "Just follow the stars and you won't fall."
          The words spilled from my mouth
               as we walked under the Milky Way
                  on that starlit night
             Down a mountain-side to a stream
          Where the Stars shone brightly
     and the Chi of the Earth roared strongly
in that lack of Moonlight

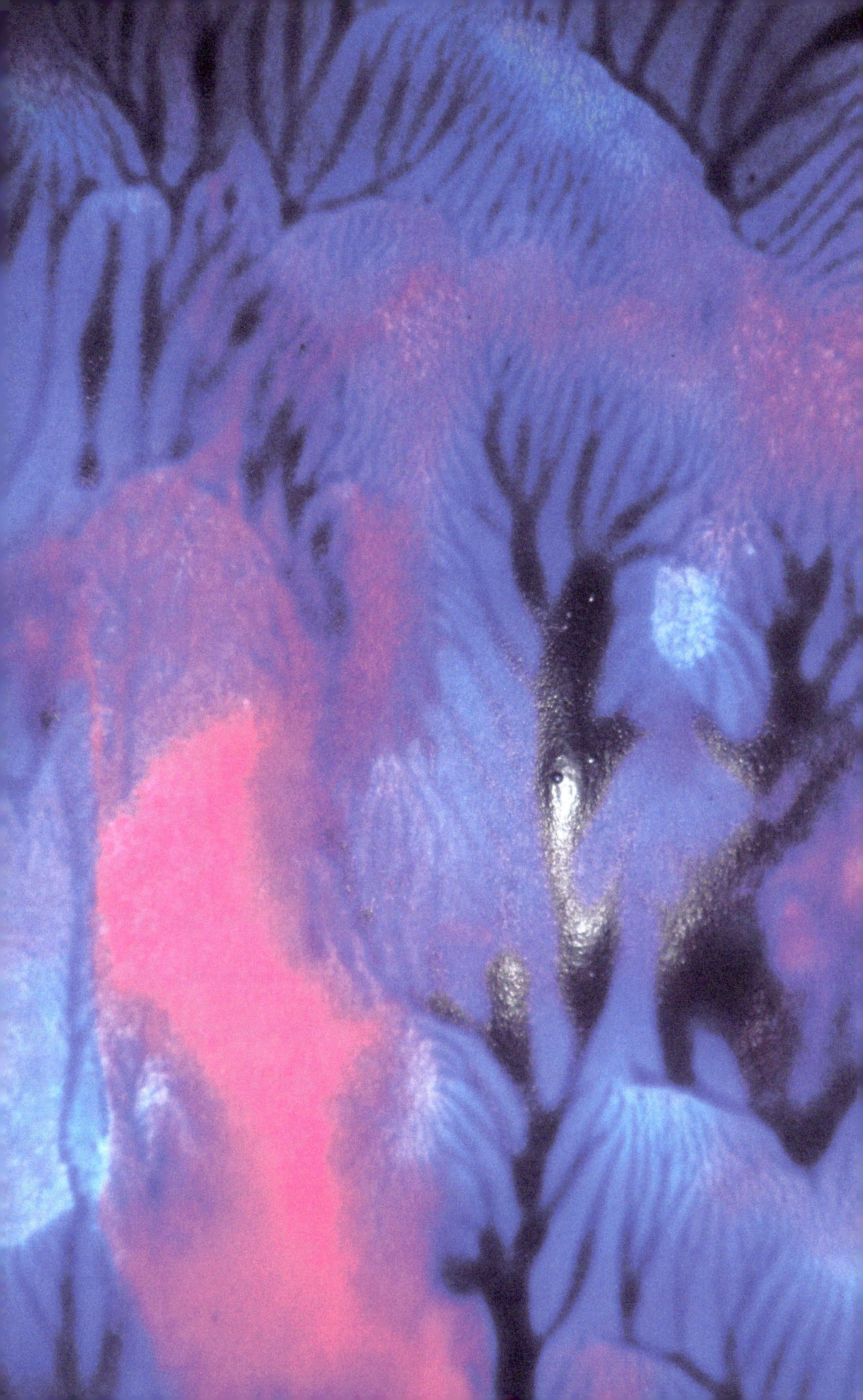

Solar Storm

Earth changes,

Solar storm wails.

        Silent storm goes through me

            Raging Electric.

        I vibrate with solar protons raising my vibration

            Connecting me to the Universe,

                Connecting me to you

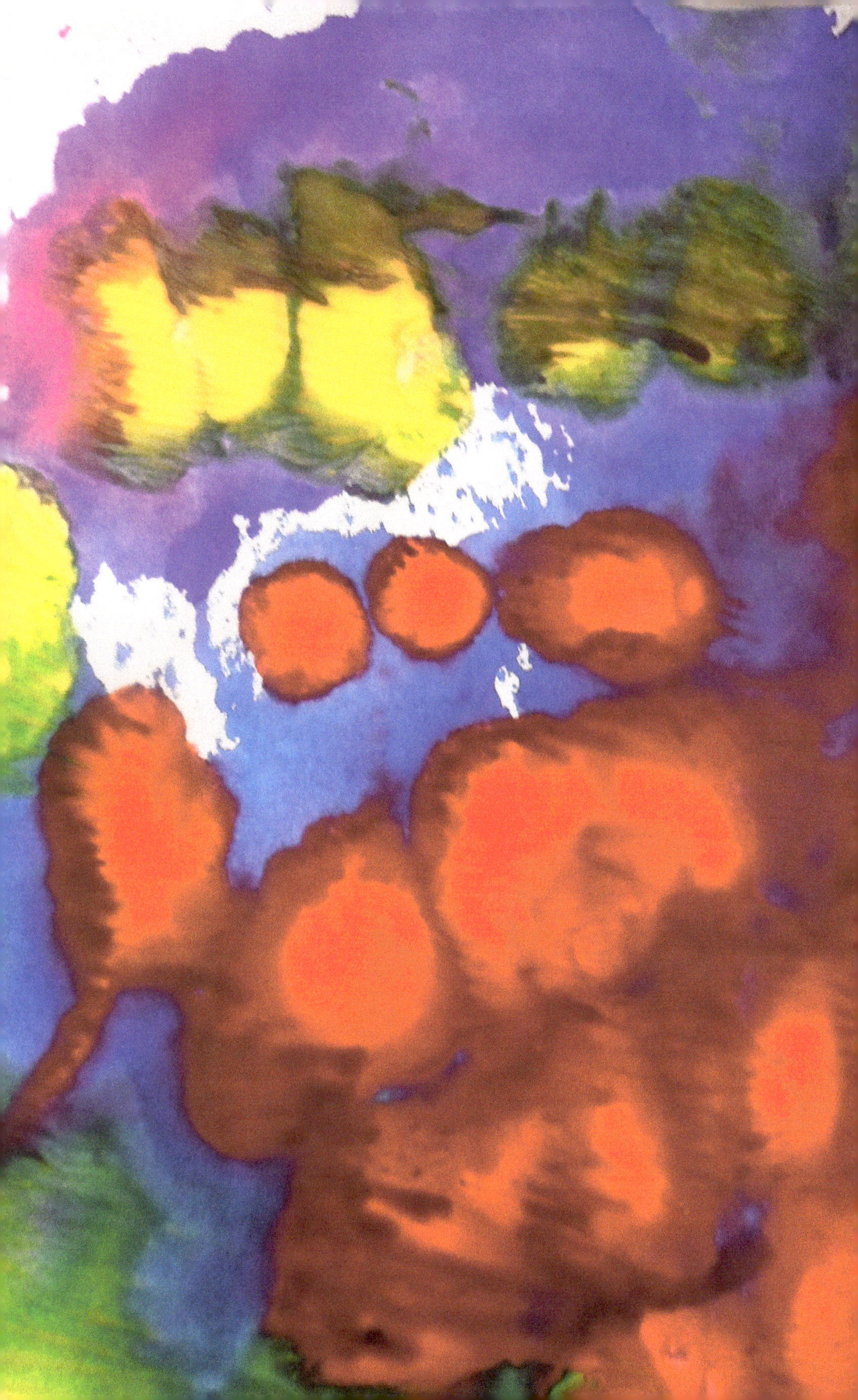

Deep Cold Fog

I am freezing into a stone statue of a woman once alive
Fluid movement is gone from my body.

My body is wracked with pain, my hips are wounded
My Heart is wounded,
My eyes are swollen by this river of Pain.

I have lived outside my body for too long,
Now when I step back, inside is the unbearable heaviness
In my limbs
In my chest
I want shelter from this ceaseless cold rain.
The pain keeps me awake
As I lay here on warm, soft sheets
My existence dulled
By a foggy cloud designed to protect me,
but only keeping me in the cold.

I need red sunsets and starry nights.
Air and Wind. Rest and Sun.

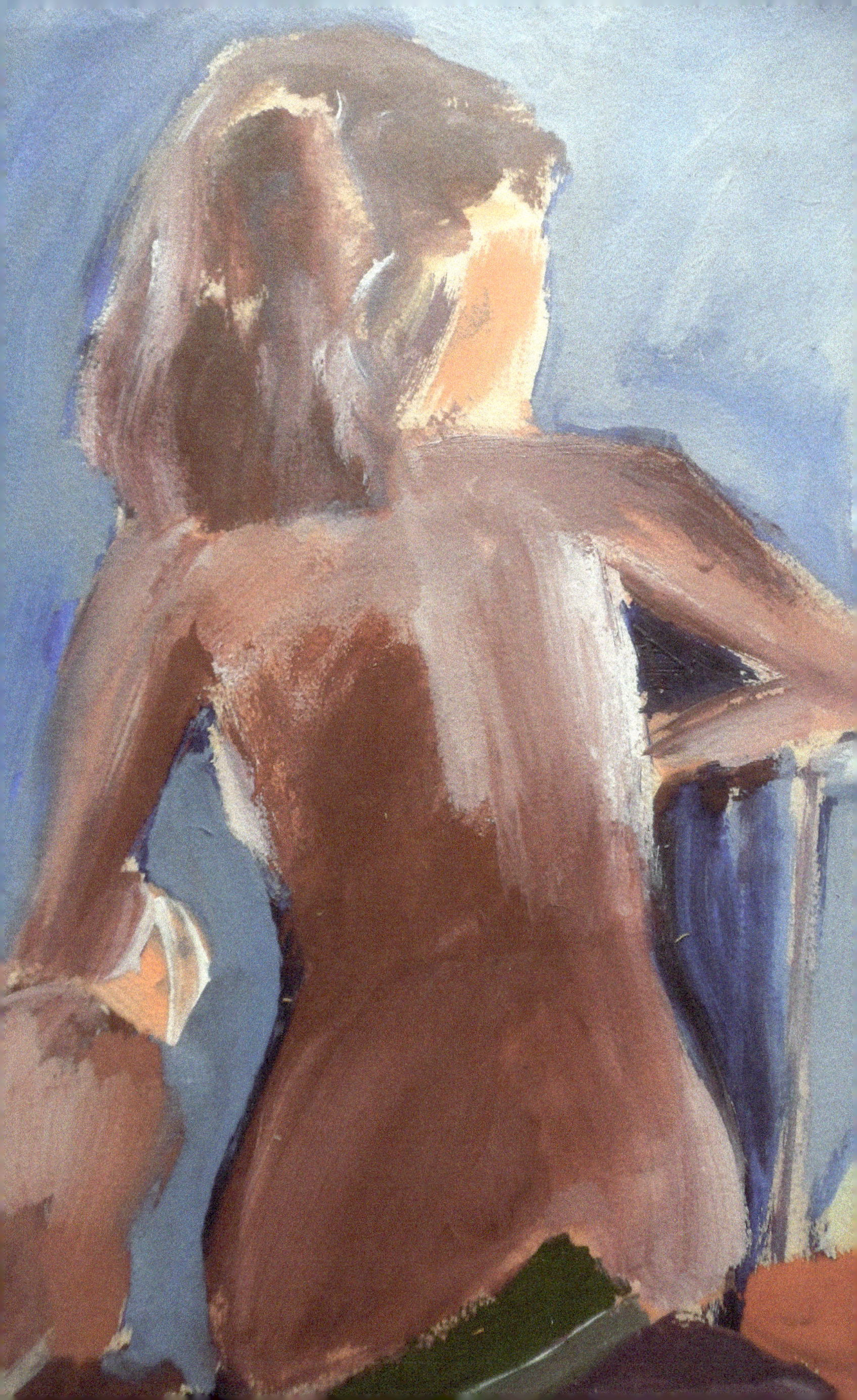

The Waves

Manicured flowers and
    Stairs of Ancient Rock
        carve a hideaway,
           here, on the Spine of the Earth.
Where everywhere green hides the life of small birds.
    Under palms full of wind, pelicans flock around.
The Chi of the Earth rages there,
    where the rock has moved straight up
            Toward the heavens.
Worn into beautiful shards and
    Sung to by the Oceans, until even the hardest pieces
        fall to sleep and are turned into soft beaches
           by the ceaseless rocking of the waves.

The waves on my beach are loud and strong,
        Here, I charge myself in the Sun.

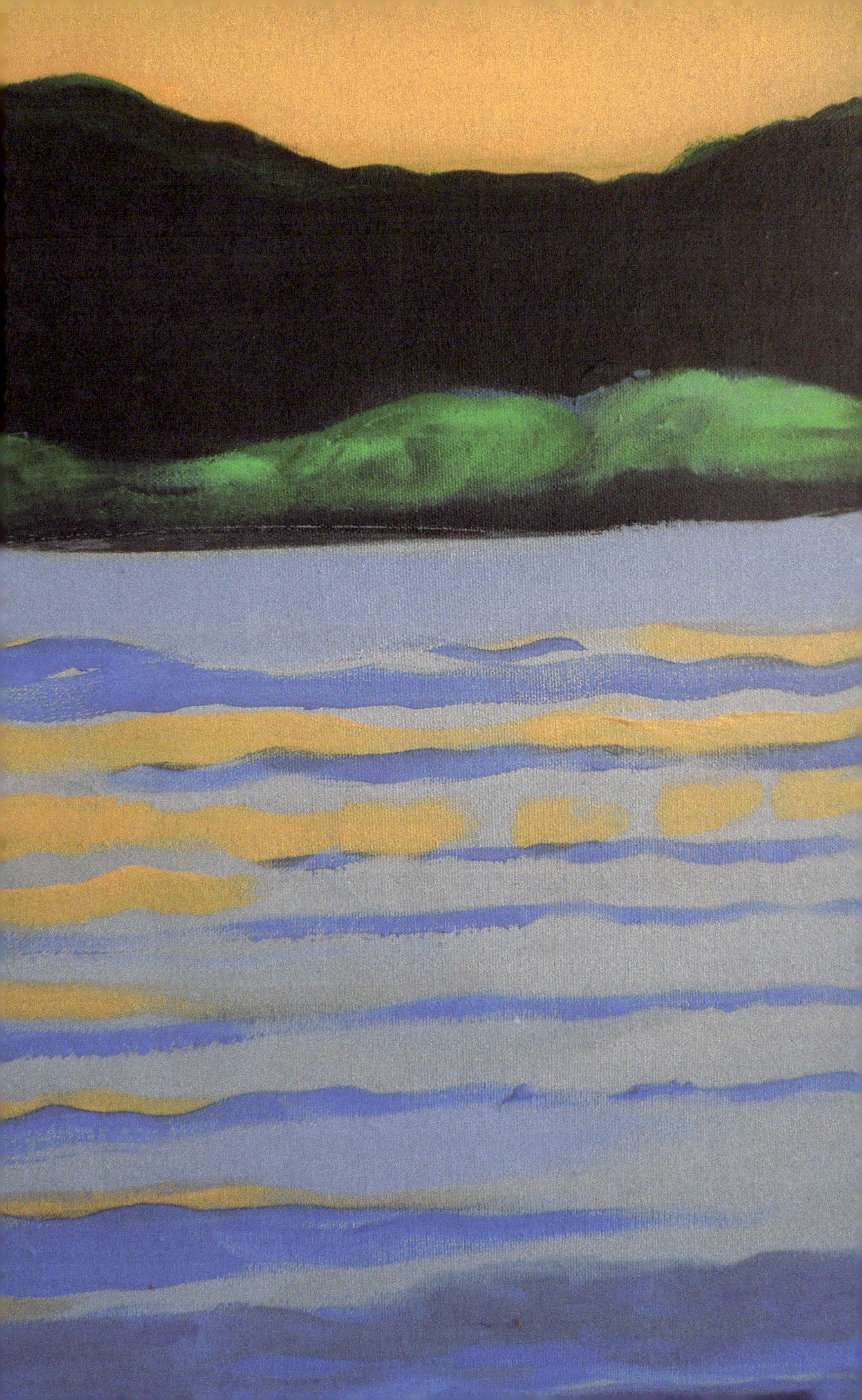

Solar Dream

In this moment,
Your Spirit feels so close to me.
Tonight, I will indulge in
the luxury of dreaming of you.

Perhaps I will find you and
You will allow me to wrap you in my energy.

I am relearning the past and
You are strong in my oldest vibrations.

You arrived to make me remember
I want to live again,
To live in the fire of your arms again,
To arm the fire of your soul, and then
Ignite your dreams.

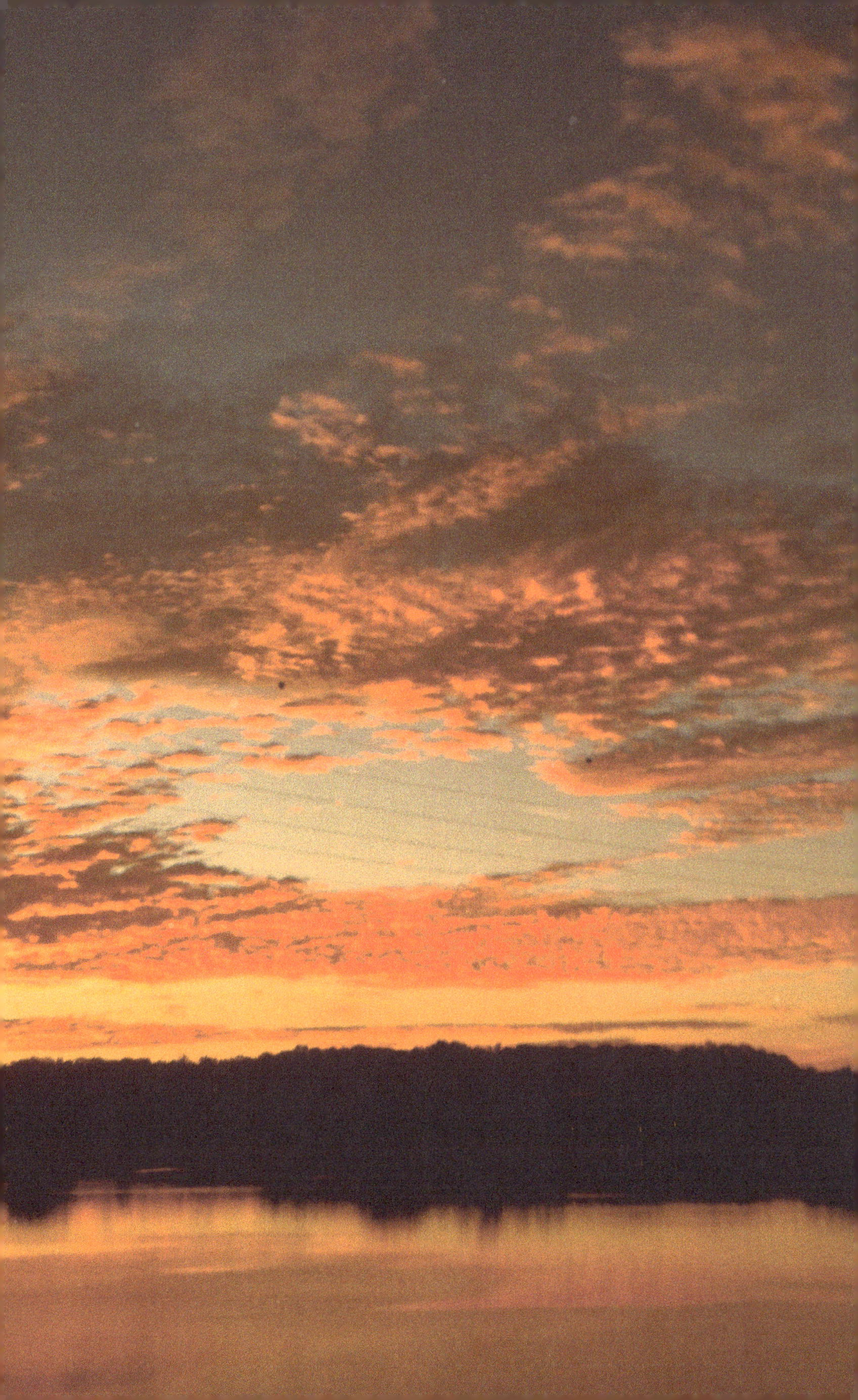

Safety in the Storm

Every night I go to the place where I find you.
As I fall asleep,
you guide me to my dreams
and watch me waken safely.

I return from a celestial safehouse to find you not present,
but everywhere around me.
It took practice at first, but now I fly to you easily,
Fly to the place where I am safe with you.

I lie next to your scent
as you encircle me in an energy dream
of spiraling universal dimensions,
increasing safety and warmth.

In a muted nest of indigo and crimson, I lie with you.
In your arms, I am warm.
In your arms, I find safety in the storm.

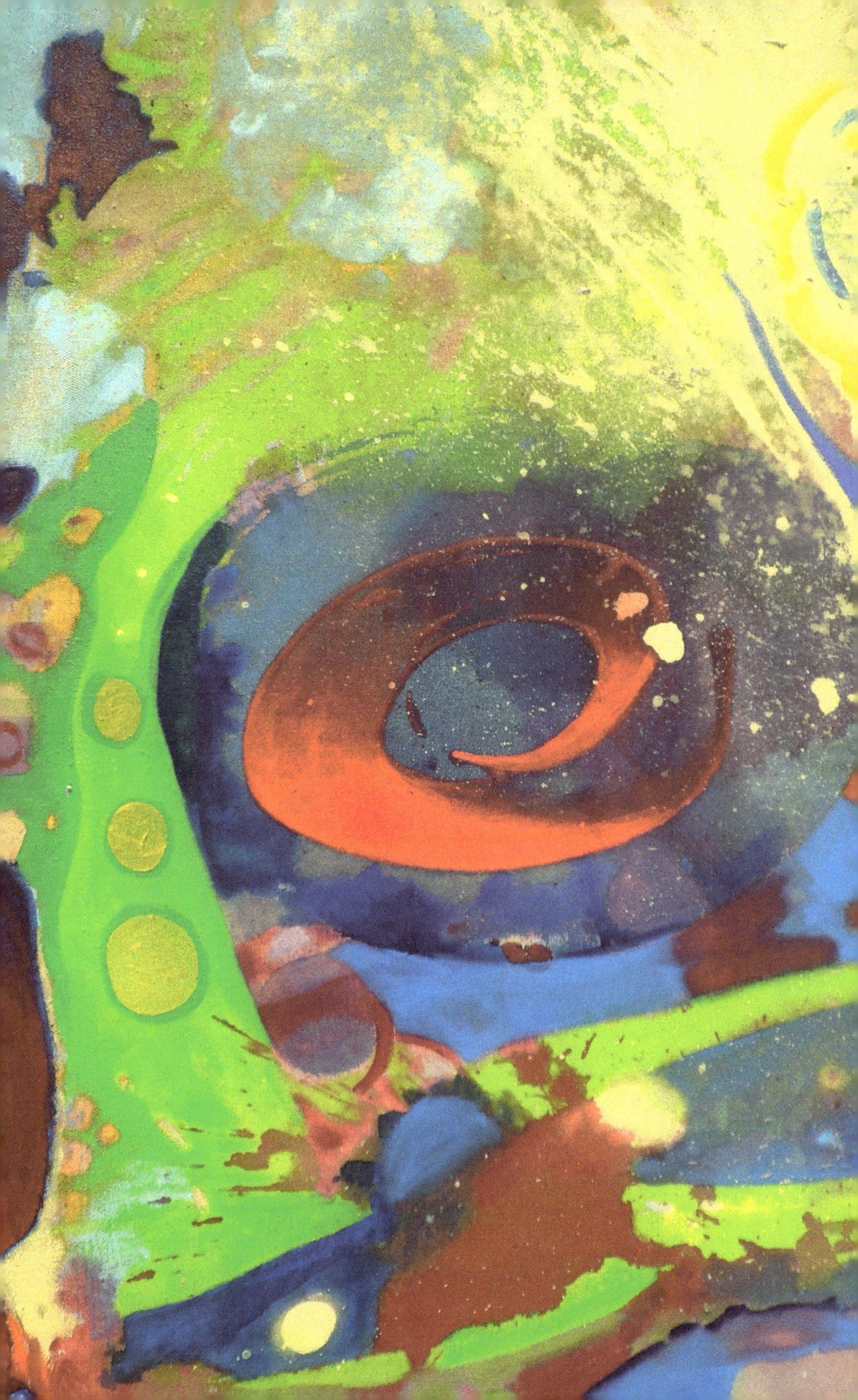

Remembering

Your visage already modeled
in the stars of Orion
I will paint you in clouds of Sage

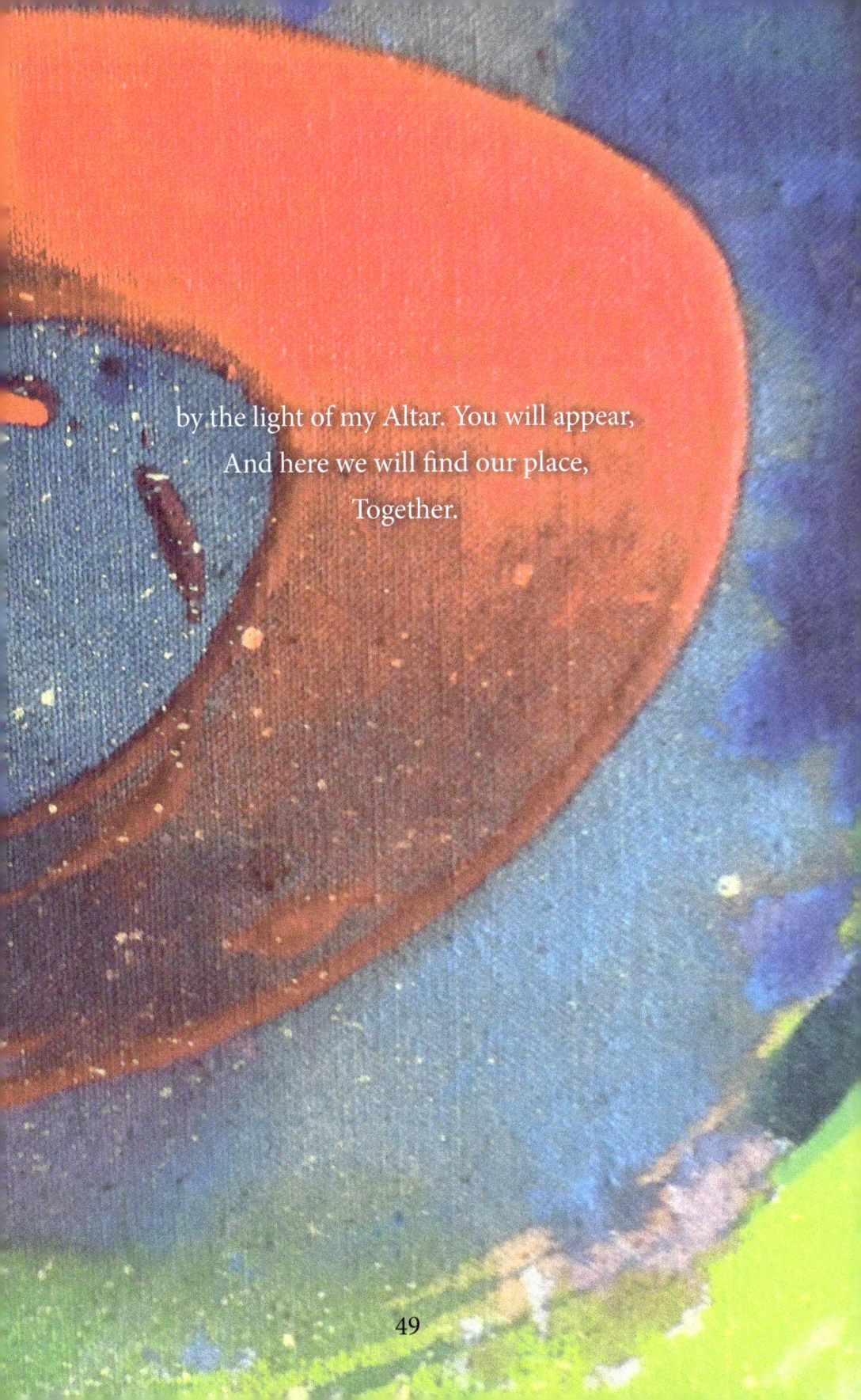

by the light of my Altar. You will appear,
And here we will find our place,
Together.

A New Moon

Gone are any days of warmth
Only my Moon lights my path, but dimly these days.
Moonlight on the Water was my name,
Still my strength,
Even as the New Moon is unseen
So it is, that my path is dim and difficult.

I walk the trek to the mountain peak in the dark,
Alone with my Guides in the starlight.

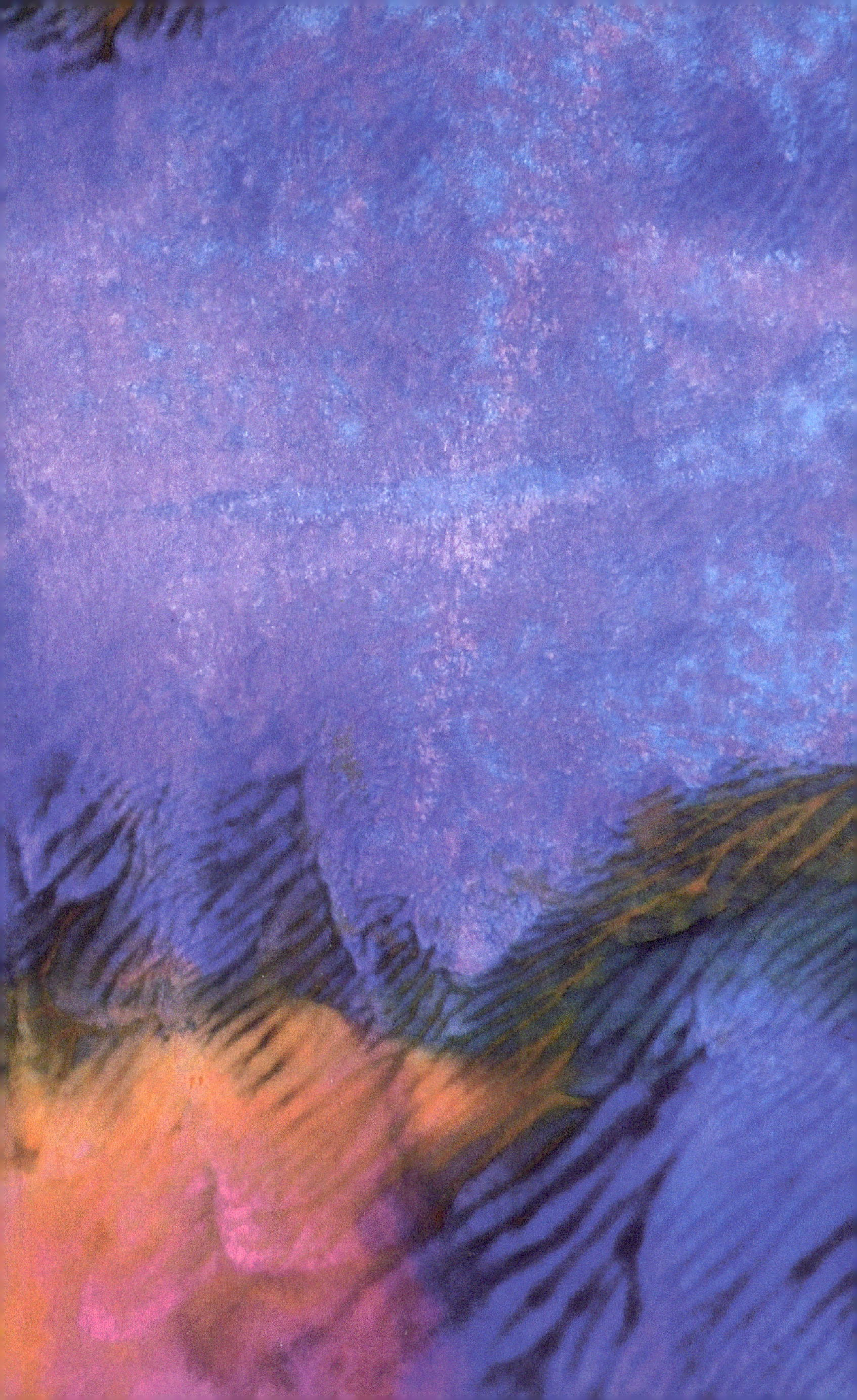

Point/Counterpoint

Sunday/ Today:
I am so full of pain that
everyone can see it on my face.
How can I get clear of this?
I am hoping this is the process
of getting clear.
All this pain being recognized

coming up from its ancient burial.

Vibrating through my body

and resonating out

Frozen in my face,

Holding in my arms, and legs, and my hips

As I move, afraid to let go,

I feel the fullness of whatever is moving through me.

(Point/Counterpoint, cont.)
Monday/Today:
I have made reservations for Barbara Zaring's painting seminar
in September, I got the last spot.
I called Pat, she made me cry.
I got info and am making plans
to travel home and to Cabo.
I wrapped soap for my sister.
Got materials to make crystal jewelry

and to use for feng shui,
Also, tools,
Batting for the quilt to make for the baby,
I ordered silk clothes and perfume
I bought two photographs
and invited John to come – sent note,
I bought nine canvas board pieces for small paintings
I made a potholder.
Life goes on.

Deep Floodwaters

I feel drowning in these waters
My clothes weighted down in stillness
But the current is moving here.

Currently,
My raft is swamped, overturned,
I am unsuited for visitors.
My injuries, unseen of light,
are stinging through me
like huge blades that cut through me
Then move through my heart,
till I can't speak, or breathe,
Much less think.

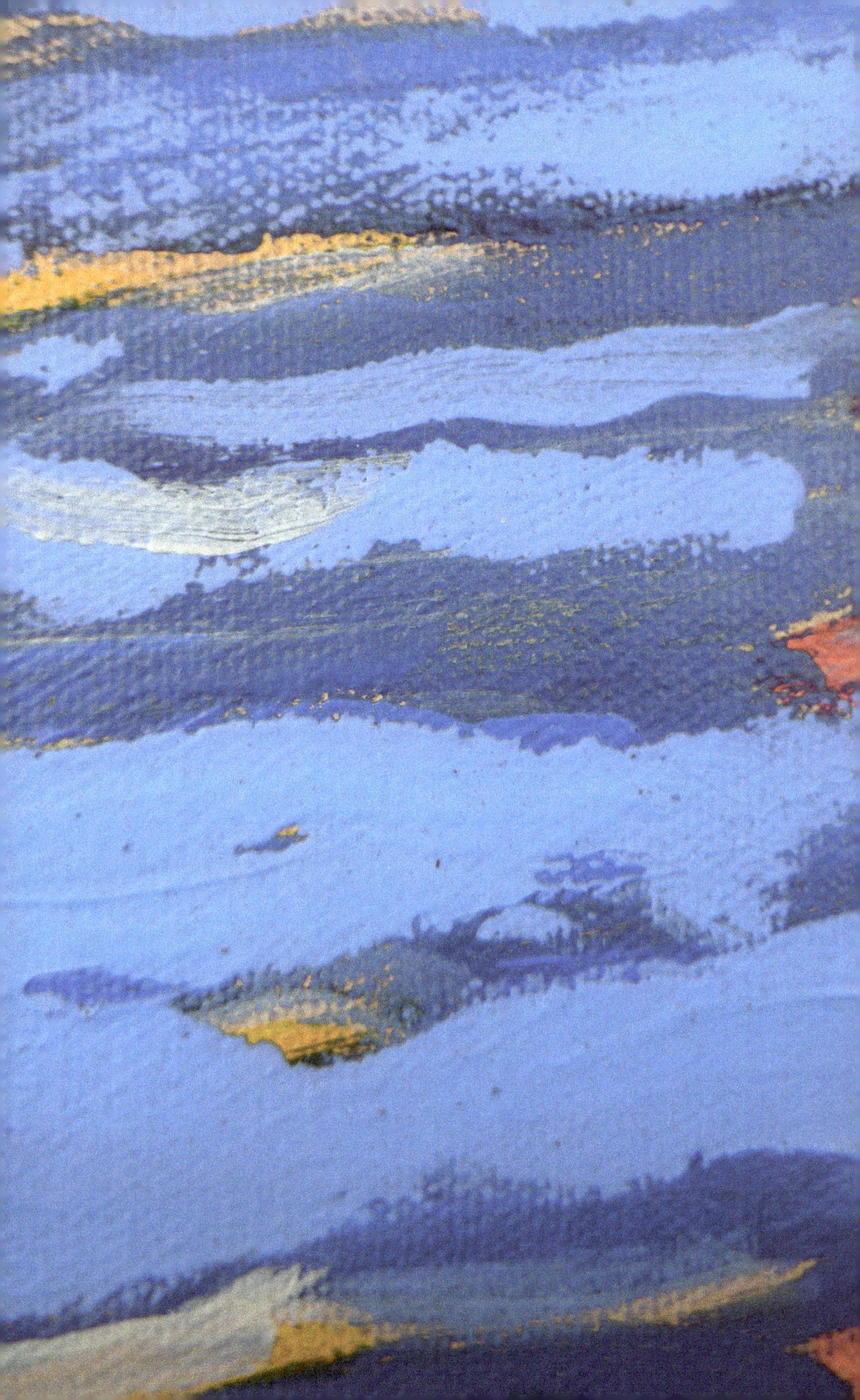

Singularity

So, are all women in the world
a collective expression
of the divine feminine?
As the One Mother
is represented by the many
differentiated female deities
and spirits all over the world?
Yes, yes we are.

All the many women in the world,
present here and now,
before us and forever to be,
reflect the totality
of the One feminine principle,
like the many facets of a beautiful jewel.
We are the many Goddesses,
We are together the One Diamond.

Lightning Bolt

The Serpent Priest
Is the Priest of the Chi,
the Kundalini,
teaching of Maya, He comes here.

A Lightning Bolt of Rainbow Energy
blasting through all physicality
Realigning, unblocking,
Strengthening.
Dissolving the Illusion
Honoring the One
Guiding me on the Path of Light

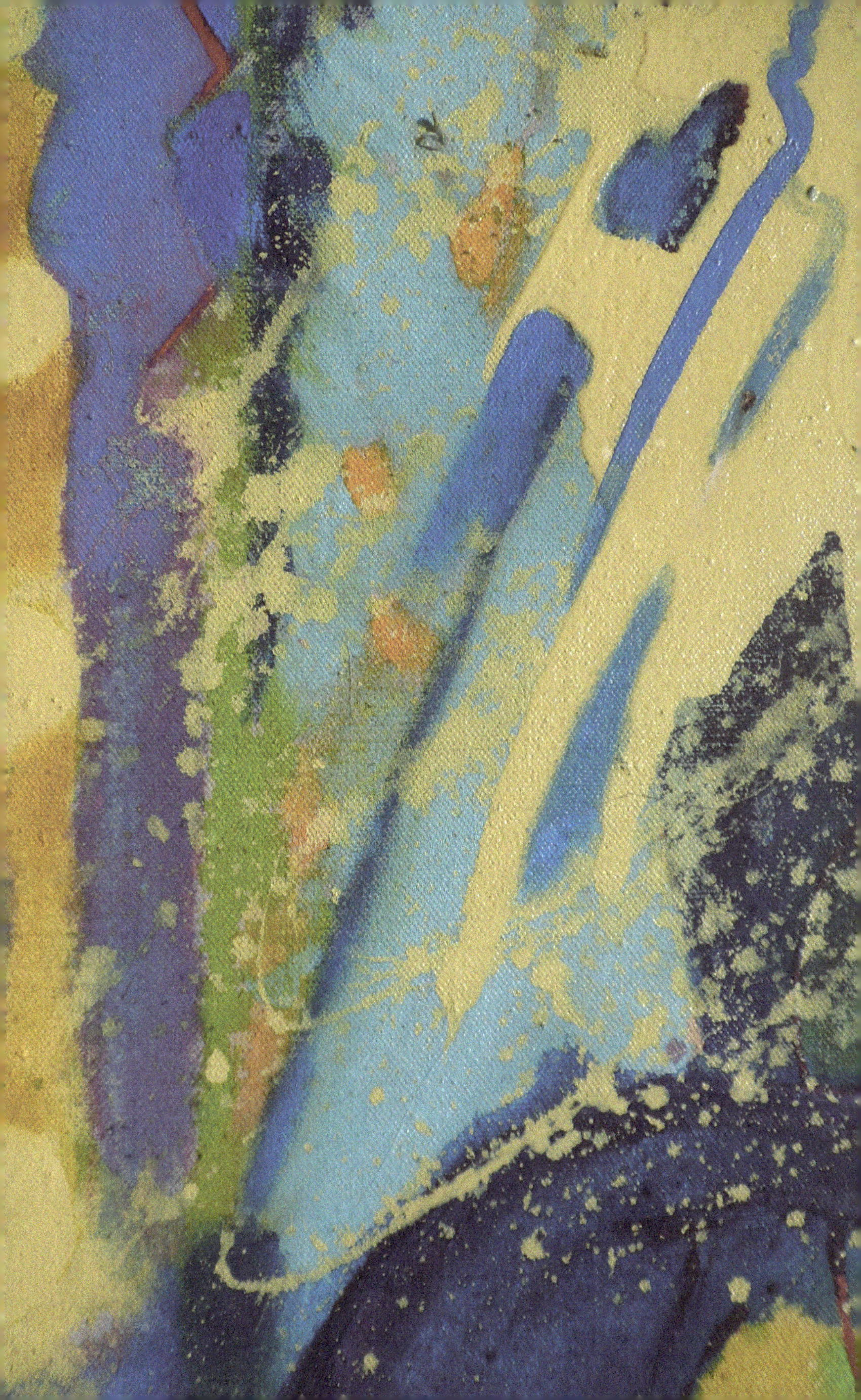

Earth Quake

I followed your advice:
And went into the Quiet.

Suddenly the aftershocks of you rushed through my body,
Ecstatic wave after wave.
I was transported back to the cave
where I was your Goddess and you my Priest, my Orion.

You came to me and
I gave you my intuition.
You spent weeks loving me, keeping those fires
In the cave of the Oracle,
Where only you the High Priest dared enter.

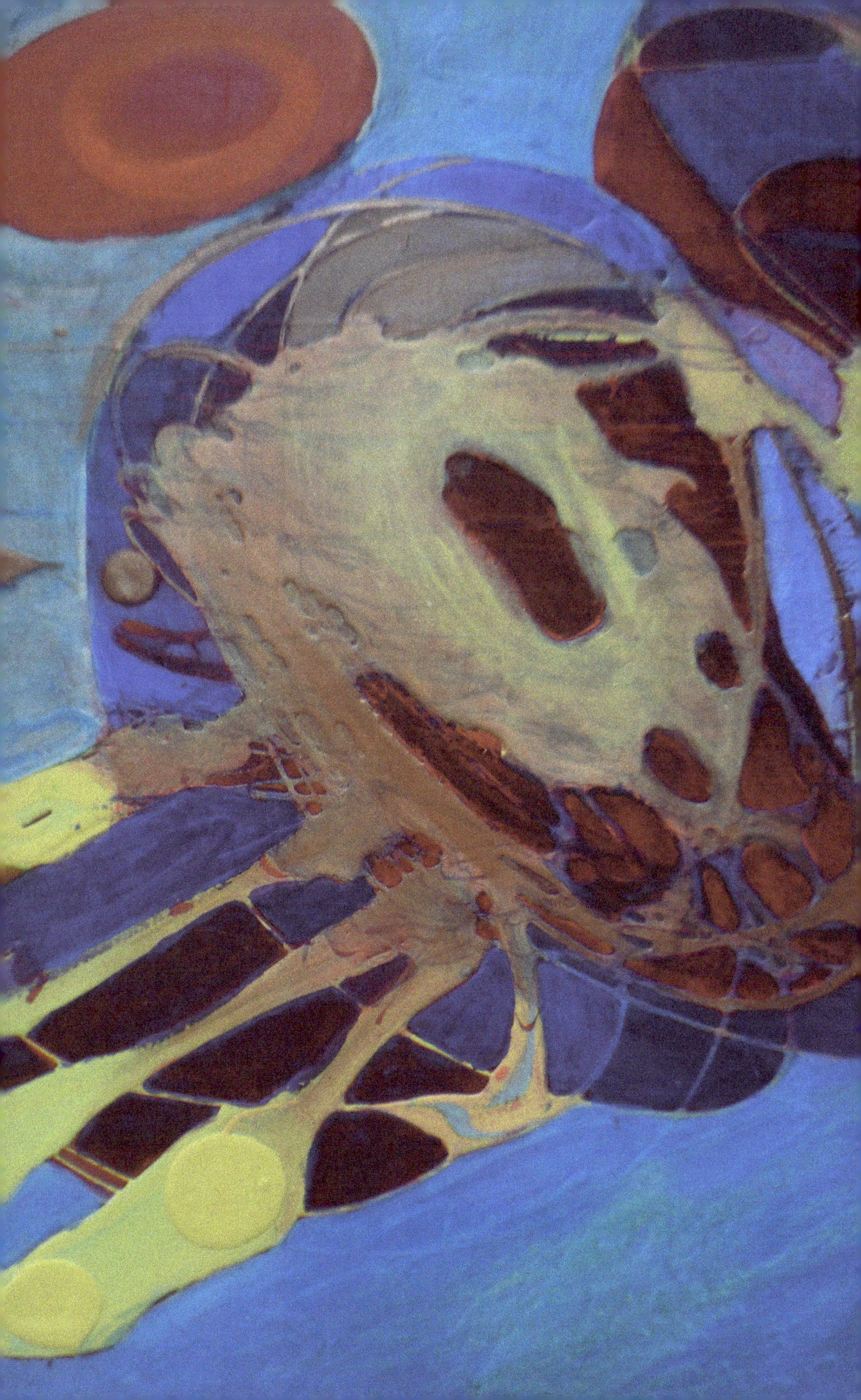

Lightning Storm

Blue Lightning, you arrived, and

I fell

Weightless

Into a pool of indigo energy
Held up by you
Swirling, silvery, hypnotically
I recall every cell in my body in ecstasy as you fed me
Every molecule realigning to your magnetic force.
Your light body in sienna shadows,
Glinting red in the darkness against the Moon.
Strong quiet Shaman
Giving me the Messages
Sent by the One.
You,
Divine Gentleness,
Are the Lightning Storm
And the Ocean's Depths.

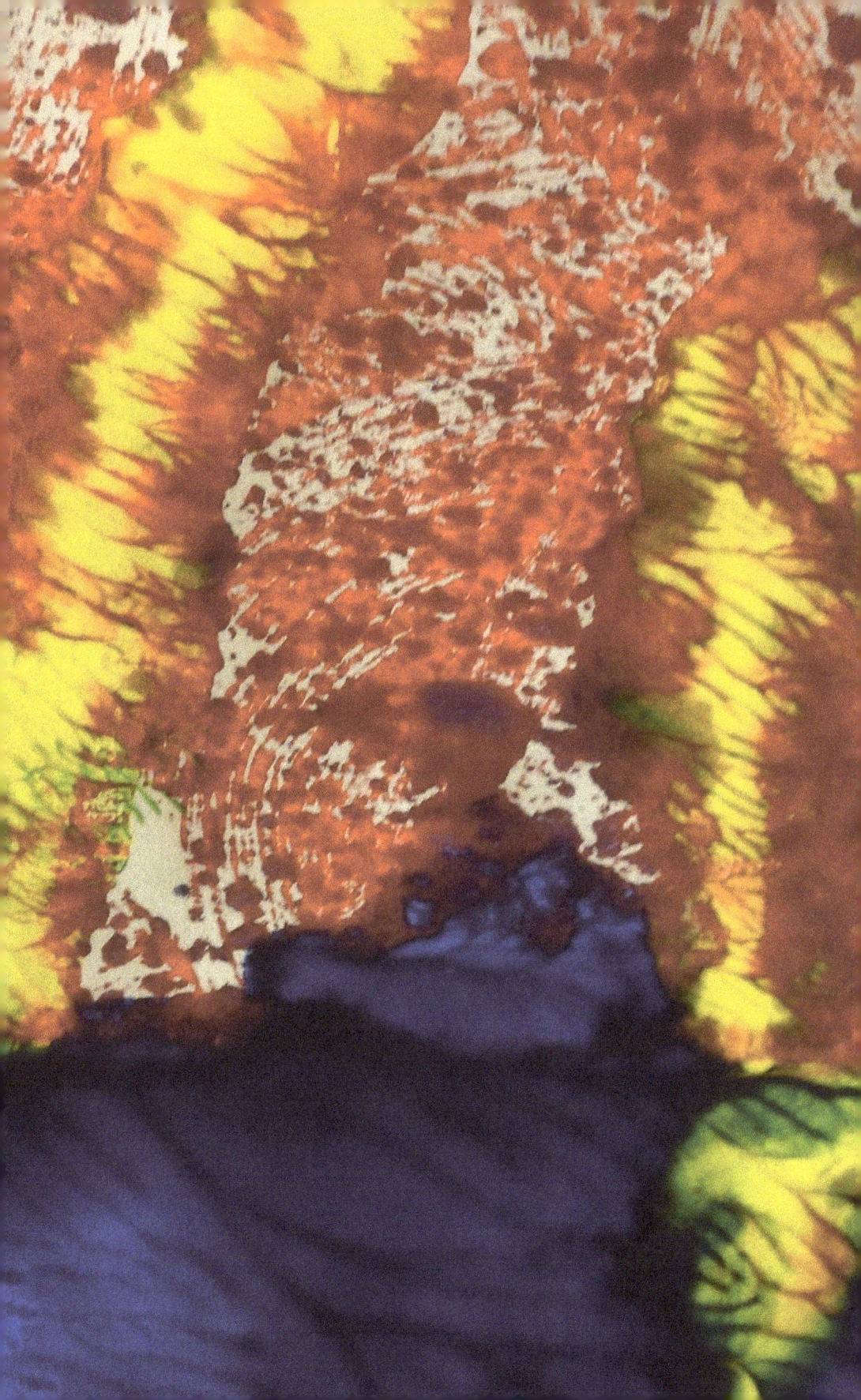

Golden Sunshine

Ice-Blue iris
One bloom above
The deep red rose,
The many roses.

The golden dust
That paints your petals
Reflects the sun
Clear sun.

Connected, collected
By green striped flesh
To the sun
Golden sun
To you.

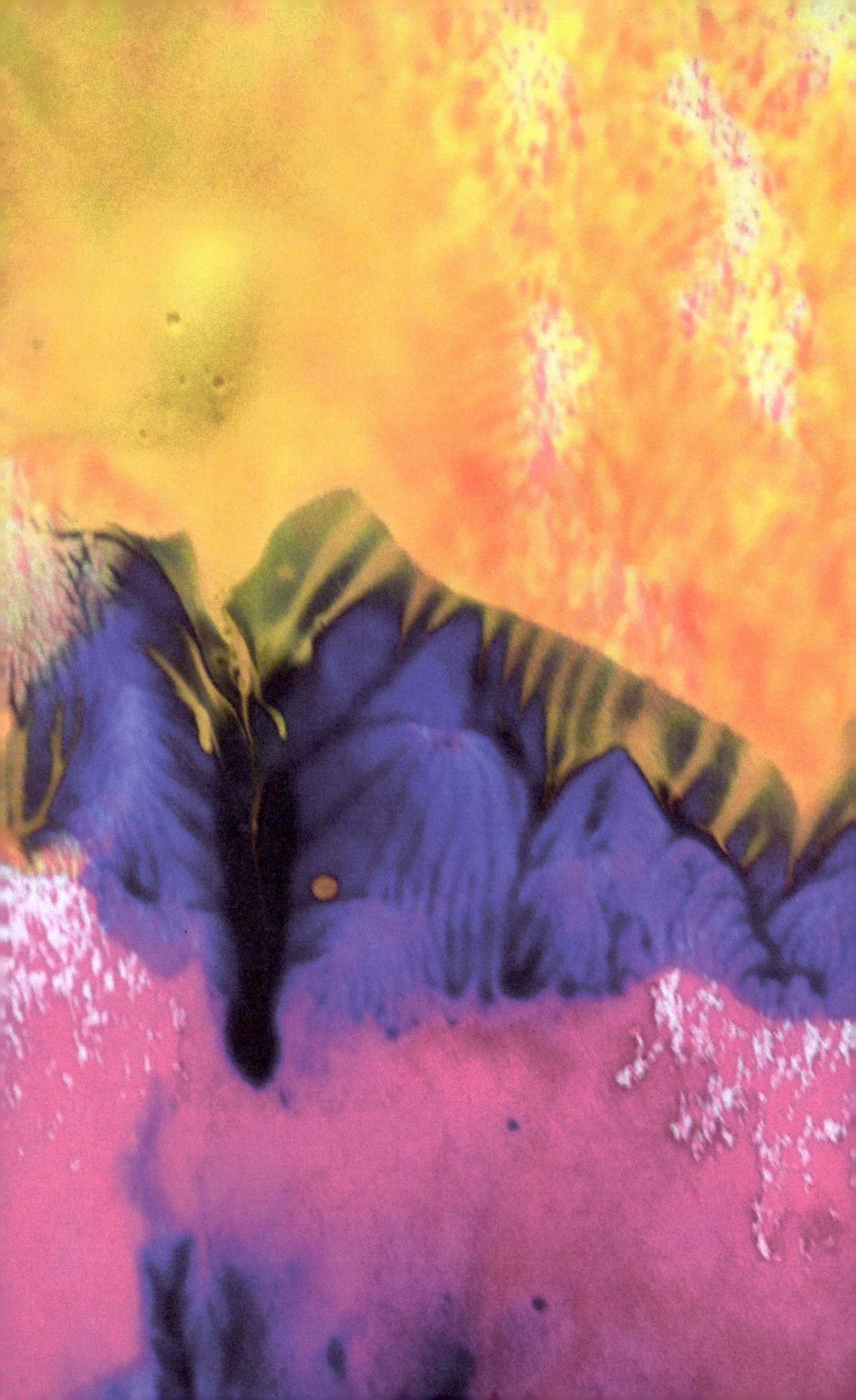

Desert Wind

You
soft grey green,
Warm yellow gold.
Your burning amber skin
Melts into my wet amorphous being
Cuts through me
Till I can't move, don't want to, only to you.

My colors are excited by your air,
Intensified as if by strong wind which purifies atmospheres.
Layers underneath, you seize me,
I am fast to you.

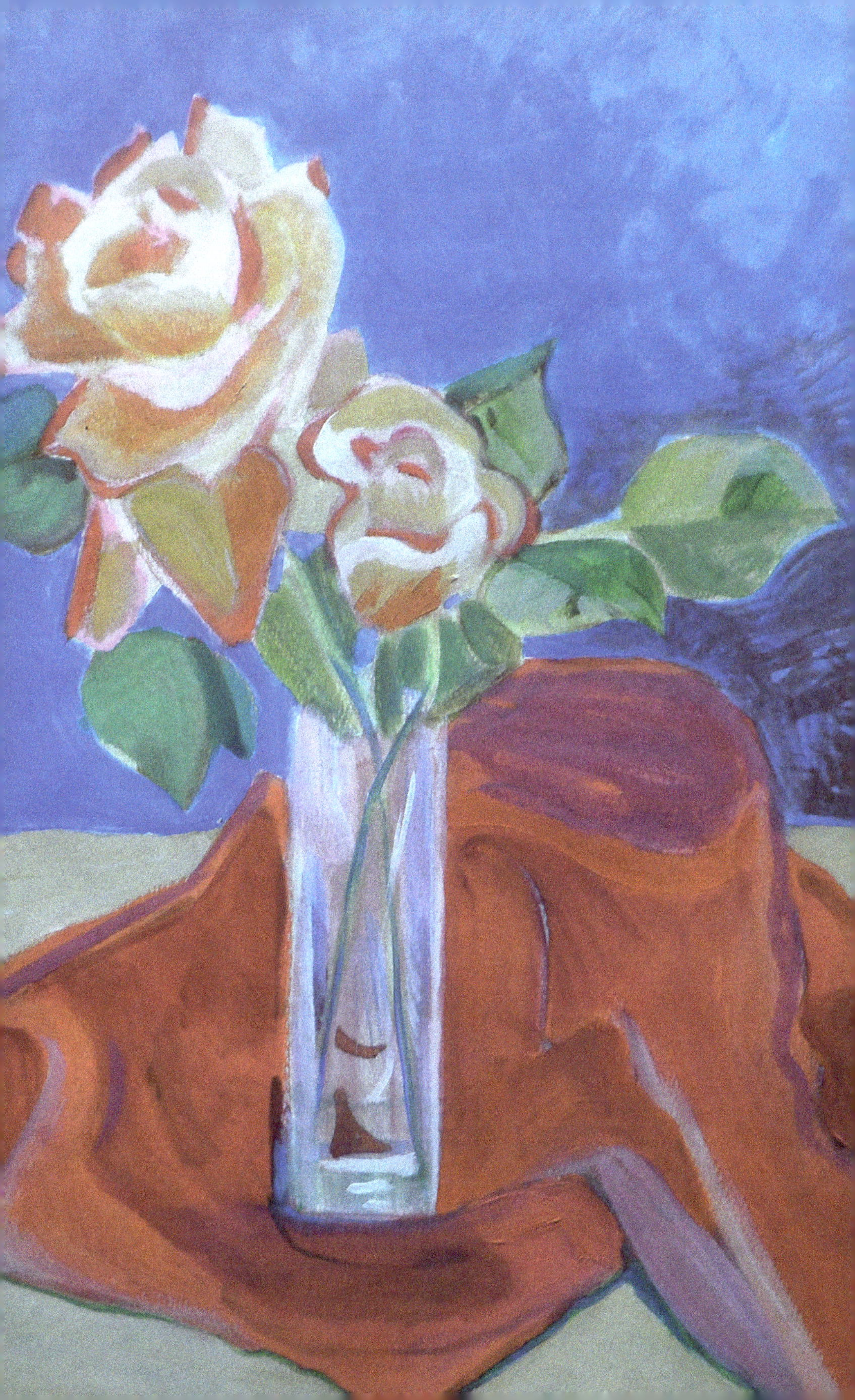

Eclipse of the Red Sun

The quiet seas we sailed to Crete:
Kreta, like Jason, sailing through the mountains
with flowing red hair, you,
Already in the stars.

Soon you sail to be a sea's length away
You would sleep alone at your camp
Leaving me to write songs
To woo you home.

Now, this eternal rain
makes the sun a childhood memory
and chills this day to its grey core.

Stop ceaseless wind
And send my red sun back to me.

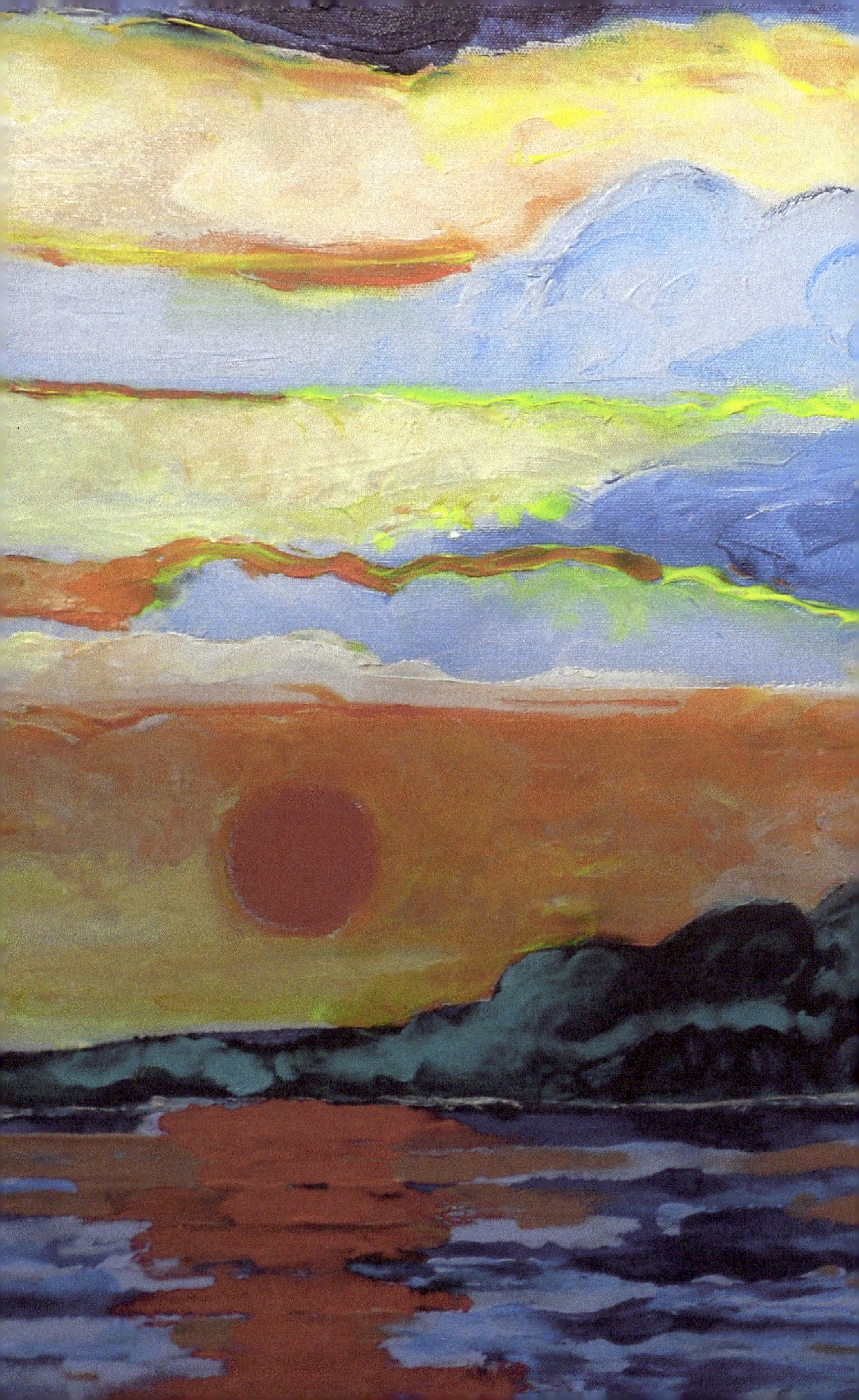

Ice Storm

Will these cold winter winds
Never end never cease?

These cold winter passions
Never warm only grief

Only grey screaming
Silence
From a far distant siren
Breaks this dream-covered island
Into Day.

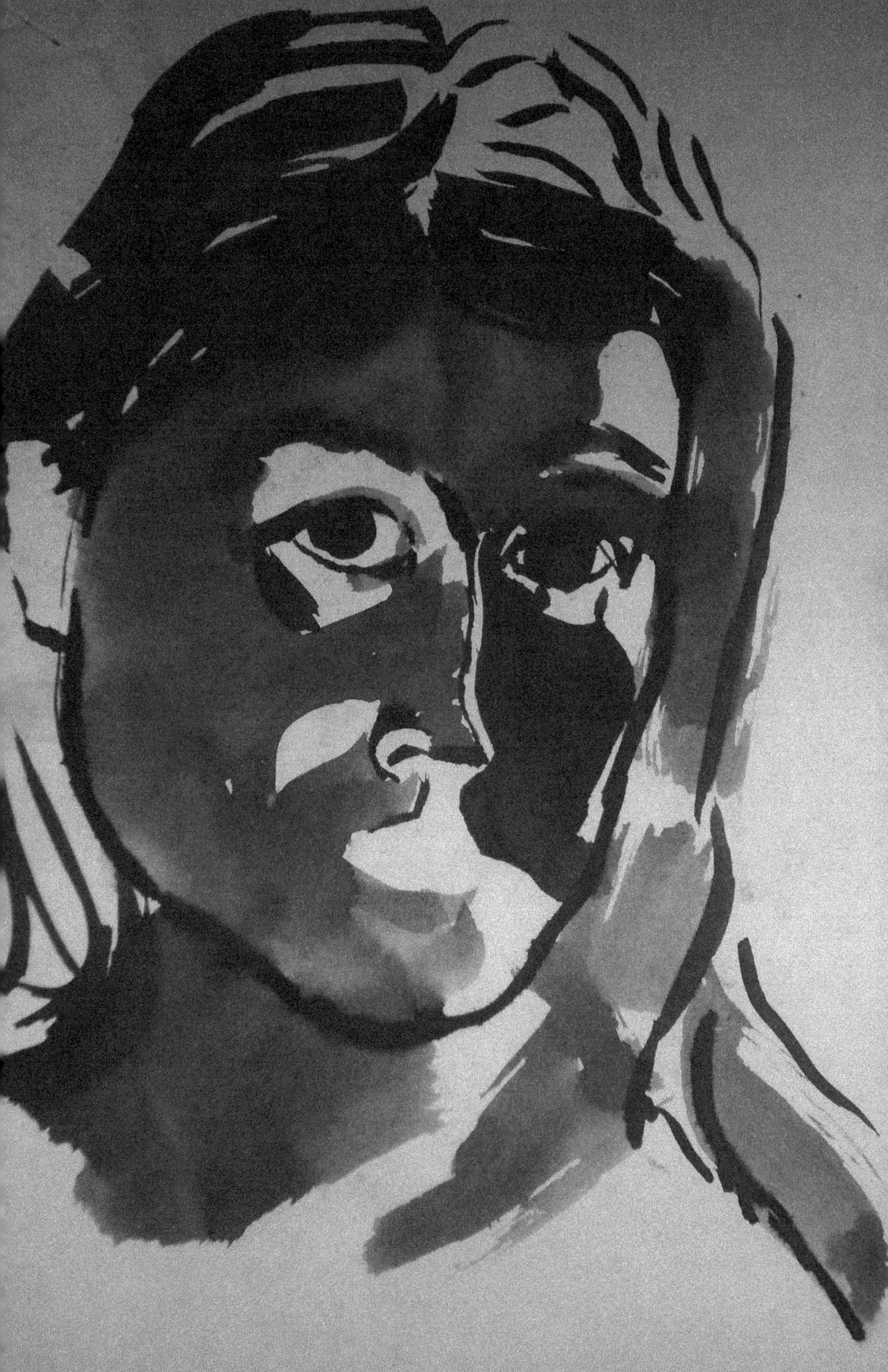

Magma

In the center of my being
You have touched me,
Come right to me
Warmly held me, wetly wooed me.

I am
touched hotly
by you wholly.
Loved far apart and closely.

You drive me
up and down
but mostly
Straight to you

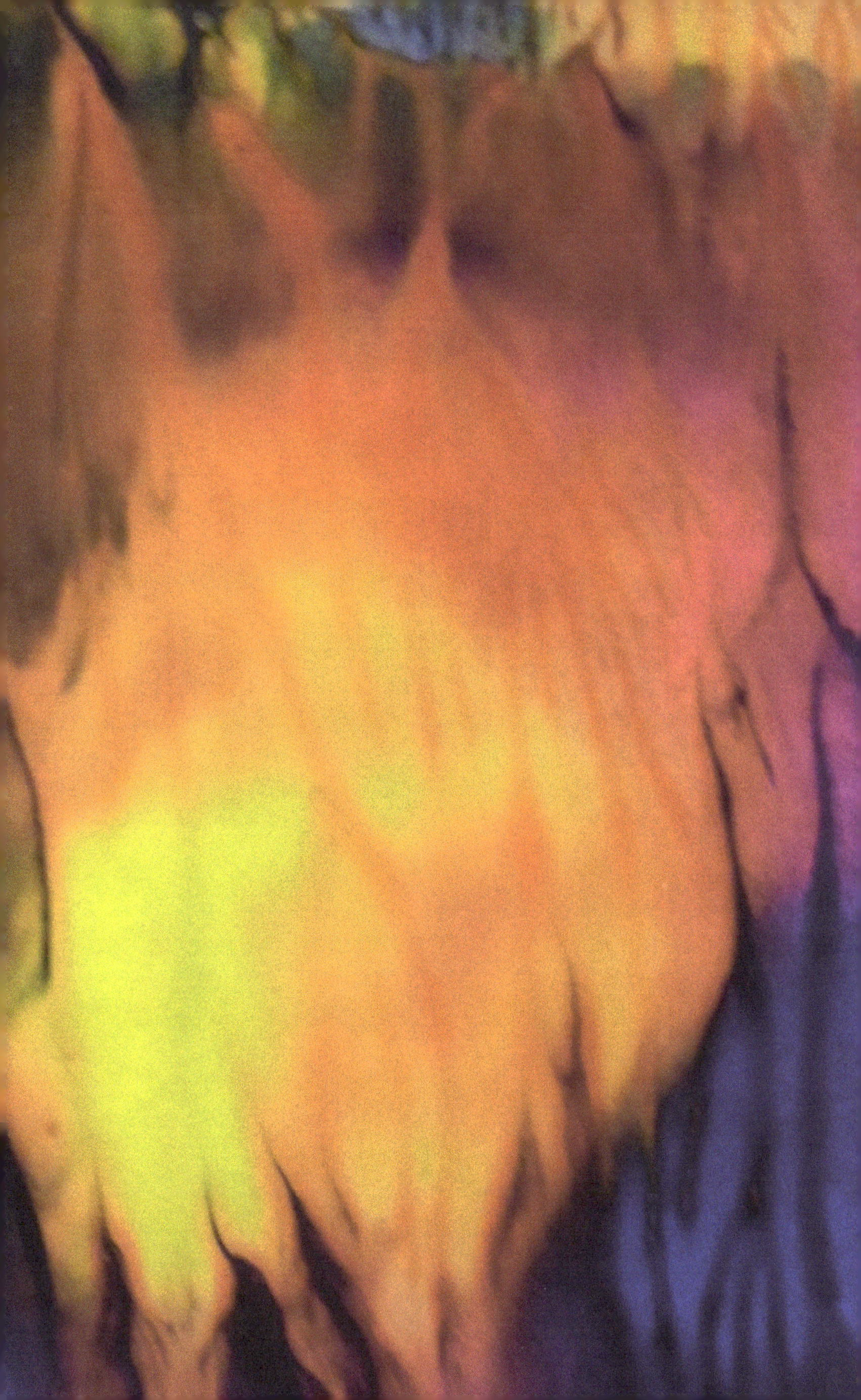

Black Hole

There is a great Emptiness
Which sucks women down.
Raging and ripping
at their senses
tearing at their
Eyes and Hearts
till they will see no more.
This deep sorrow,

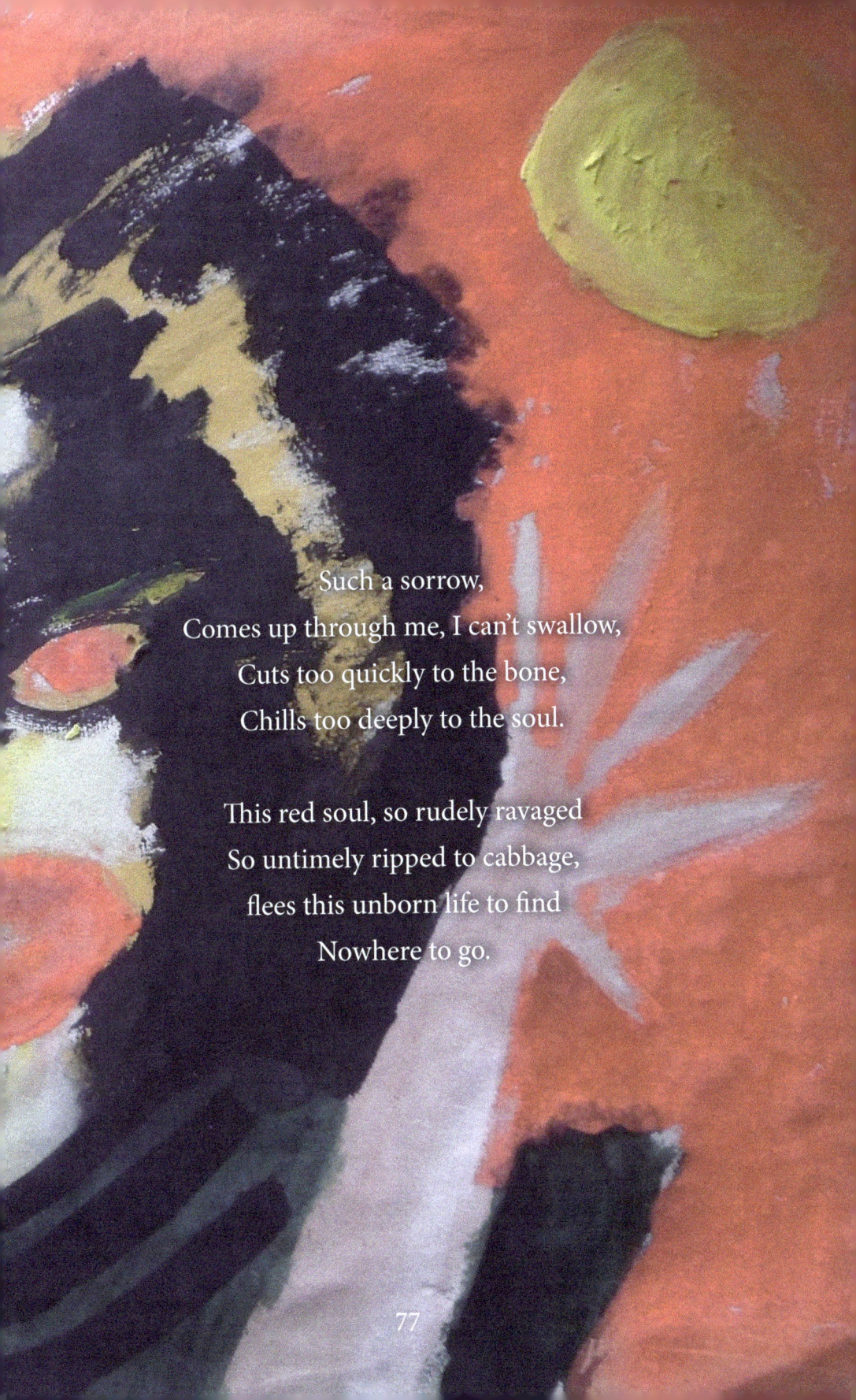

Such a sorrow,
Comes up through me, I can't swallow,
Cuts too quickly to the bone,
Chills too deeply to the soul.

This red soul, so rudely ravaged
So untimely ripped to cabbage,
flees this unborn life to find
Nowhere to go.

Deep Freeze

The deep sadness of this city
Comes back to me
Through this window

Flashing briefly
Softly sighing
Cold, cold glimpses
of slow dying.

This dark silence
Crushing silence
Sounding so much
like an evening
when the stars could shine forever
Fires this raging madness
Whether stars shine
silent or stare boldly
in the night.

My Father Is the Wind

Tonight, my father is the wind.
I hear his voice whisper in my ear
As the cool wind breezes across the river and
rustles through the maples and oaks.
I can almost see him standing at the edge of his lawn,
Looking out over his river,
Watching the fireworks

Which mark his independence
From this earthly existence.
Tonight, my father became the wind –
Blowing away the last remnants of a life well lived,
I feel his cool caress across my cheek,
as I realize that
Tonight, my Father is the Wind.

Urban Mechanization

You screech, Far away,
Scream merciless
In Pain
An acute agony brings you here.

Your scream always jests
of your seeming madness,

this stark urban landscape
always will test, seems
A mechanized madness befits you at best.
Your motorized silence, it never quite rests.
Your grey ash declining
These black birds are flying
Over deep songs of
sirens Away.

Harsh Light

Inside you,
I find no holds
Just warm consoling beauty.

You, shine stark,
Yellow bulb,
Make deep shadows
That confuse me,
Electrically, you use me.

Your own starkness
Seems to bruise you
You douse lights
It seems to soothe you.

But, at dusk, you come –
    All things blue now,
At dusk you howl –
    Right at the Moon now,

At dusk, my dawn
My witching hour,
  my wishing our
It soars and towers,
  it soothes me too.

Dream of Happiness

My beautiful cowgirl dog
Kisses me sweetly
my black cat, my familiar,
Watches over and stays near
they too love the song of the Geese

And the blossom of the flower.

They run with the birds and chase jackrabbits
Under rainbows
Along rivers of sage.

As a Comet Appears

The sky is alive with organic movement of celestial orbs

Early and late, I am

Mesmerized

I am electrified

by this new light in the Sky.

A great tail of a star

A million miles long

Blazes through this starry canopy of blue night
Lighting our Path on the Way
As we walk
Great Spirit watches over us in
Our Valley of Small Miracles
As a Comet appears.

Searching for Crystals

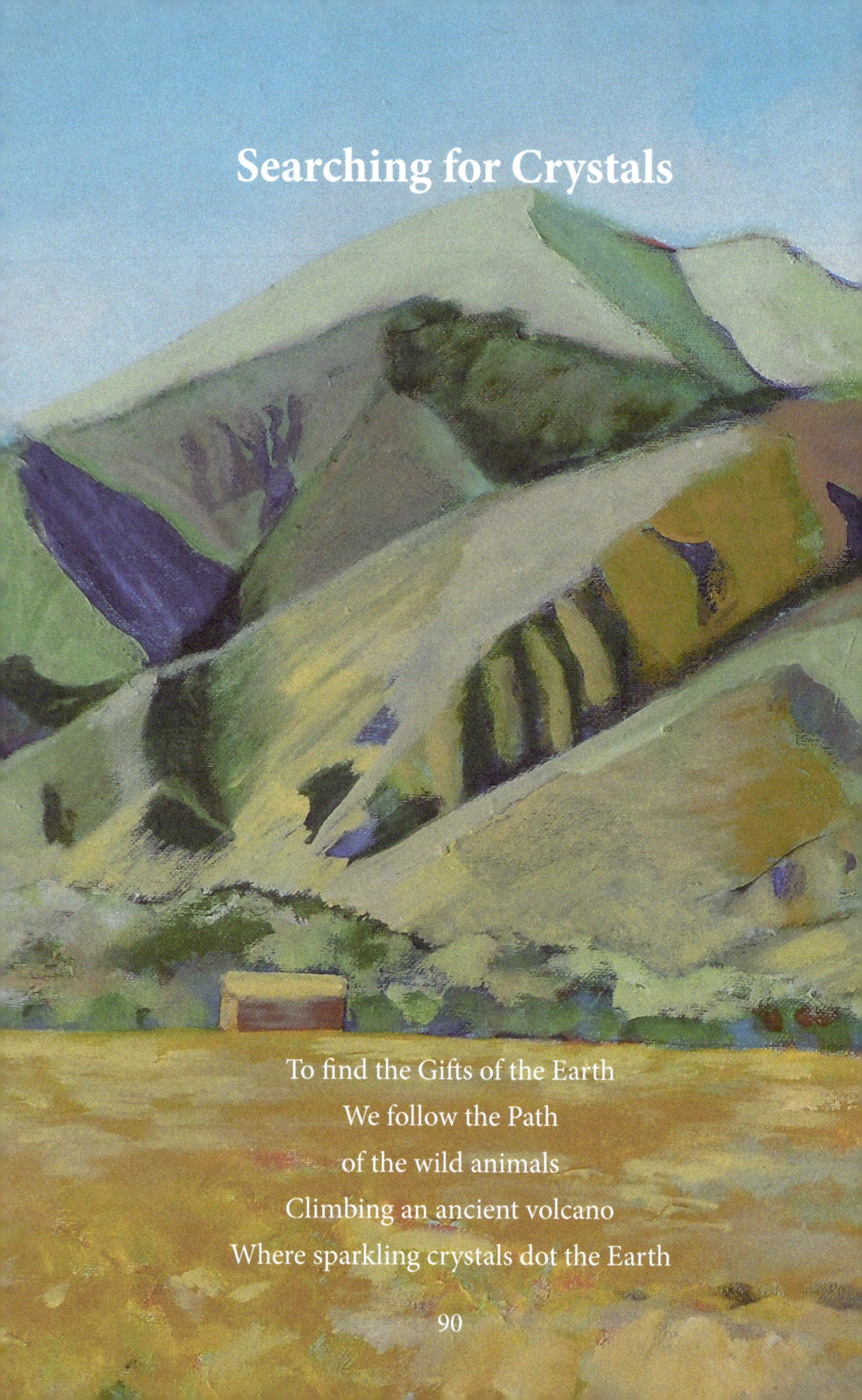

To find the Gifts of the Earth
We follow the Path
of the wild animals
Climbing an ancient volcano
Where sparkling crystals dot the Earth

We gather these Gifts
As Sunbeams send
solar waves through us,
the Light breaks into Colours
And heals us on the mountainside.

To Great Spirit

I love you with all my heart
All my being
With what Spirit I have
With every breath I take
Within every cell of my body
Within every atom of my being,
Within every photon of my aura,
I pray to be guided to serve you.
To serve you always in all ways,
You my teacher
My guide
My God by whom
I measure all others
I love you forever
I love you.

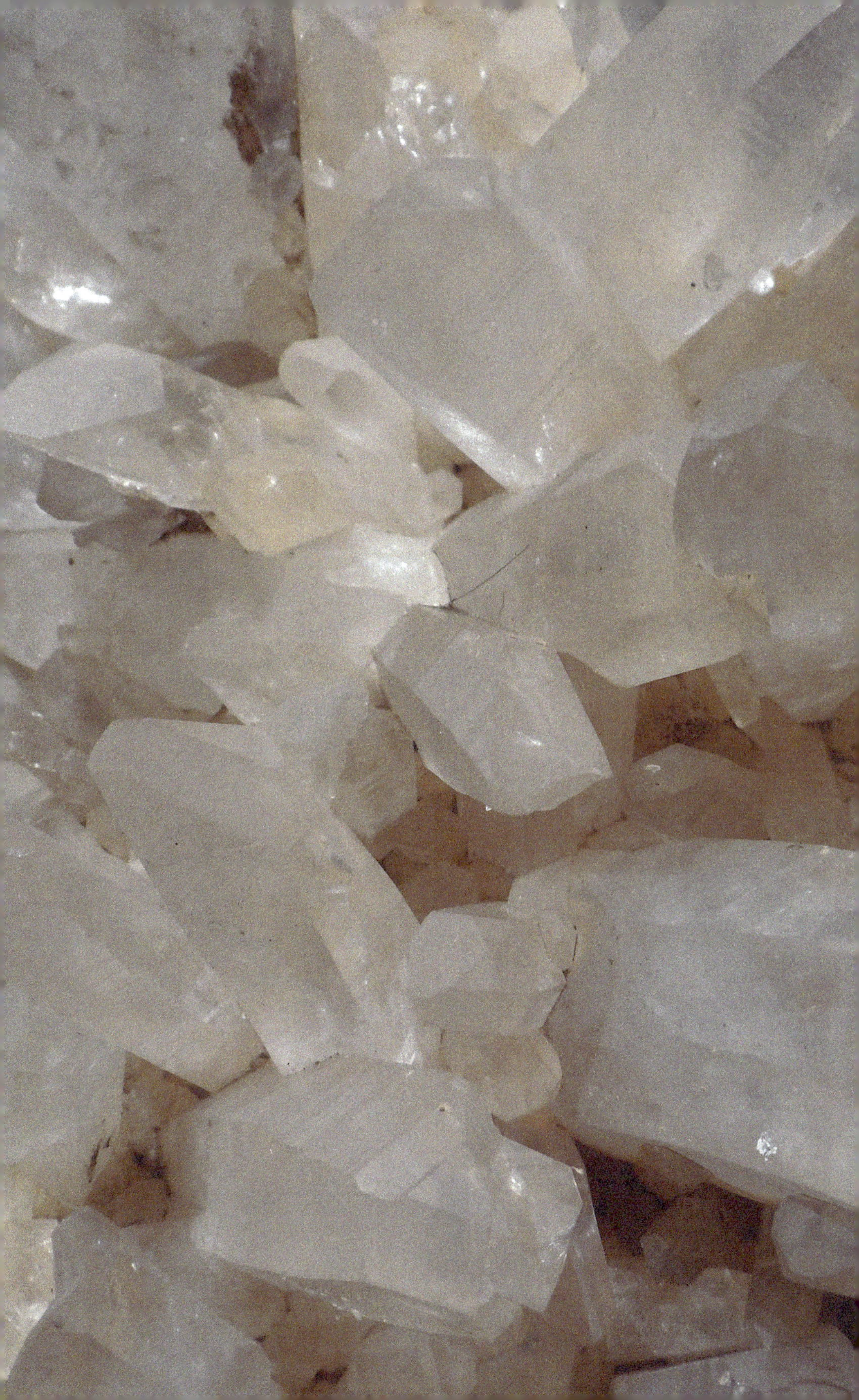

Moment to Moment

In front of an orange Sunset
On a beautiful wild mountainside
We sit together.

Twilight brings a comet into view
And I remember our past –
Picnics on an Italian hillside
With wine and olives, oranges and cheese.

Tonight, you will pour my wine
And I will look into the eyes I have loved forever.

As a storm sweeps the horizon,
You bless me and keep me awhile.
Moment to moment.

These are the moments that endure through time
Drinking each other in,
Sharing food and wine
Looking into eyes, souls,
Creating, living, loving.

Mid-Flight Standstill
(The Hummingbird rarely stands still in mid-flight)

Tonight you blessed me,
Graced me with your Spirit
What a delicious breath of you I was allowed
A beautifully balanced and most complex flavor.
For a moment,

You allowed me to inhale your elusive Spirit
To indulge in your essence
How gracious of you
To stand still in mid-flight,
I am not finished with you.

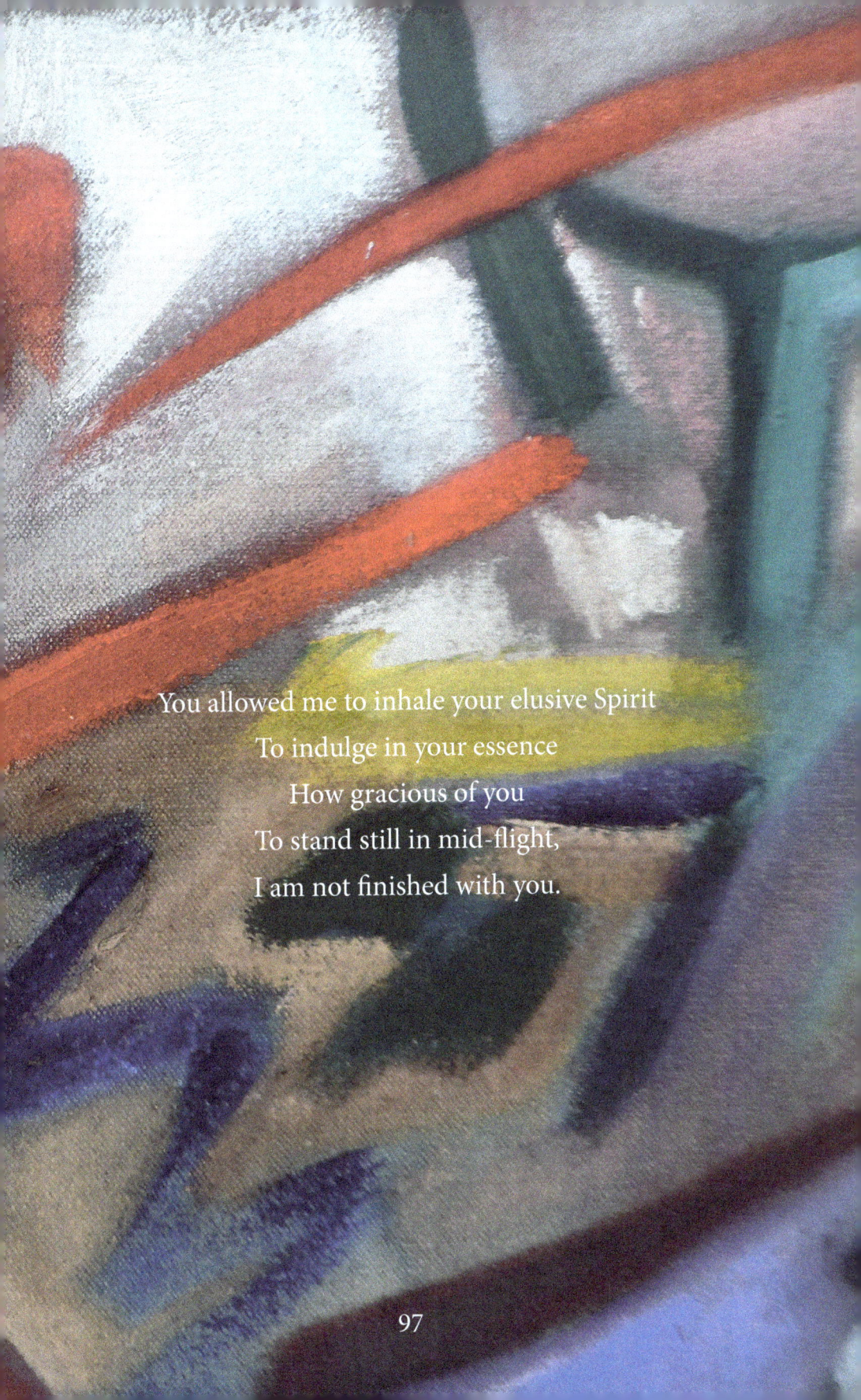

Crisp Morning Air

Flocks of Geese fly to the river that runs through this Valley
Blanca rises above the white blanket of fog
hanging over the fields
Layers of Air and Water stratify before the Mountain

Where on this Earth, sounding among the early risers,
Are the songs of Geese
Carried on crisp morning air.

Solar Expansion

You will no longer be in my thoughts.
I am releasing you.
I take back the power I gave you.

I will no longer feel ugly for you.
I no longer desire you.
You are not strong or a healer.
You cannot love because you do not know love.

I hereby withdraw my aura
From all our interrelated activities.
I wanted to stand in your light a moment
But you are standing in fear.
You believe there are limits to love.

Love is expanding.

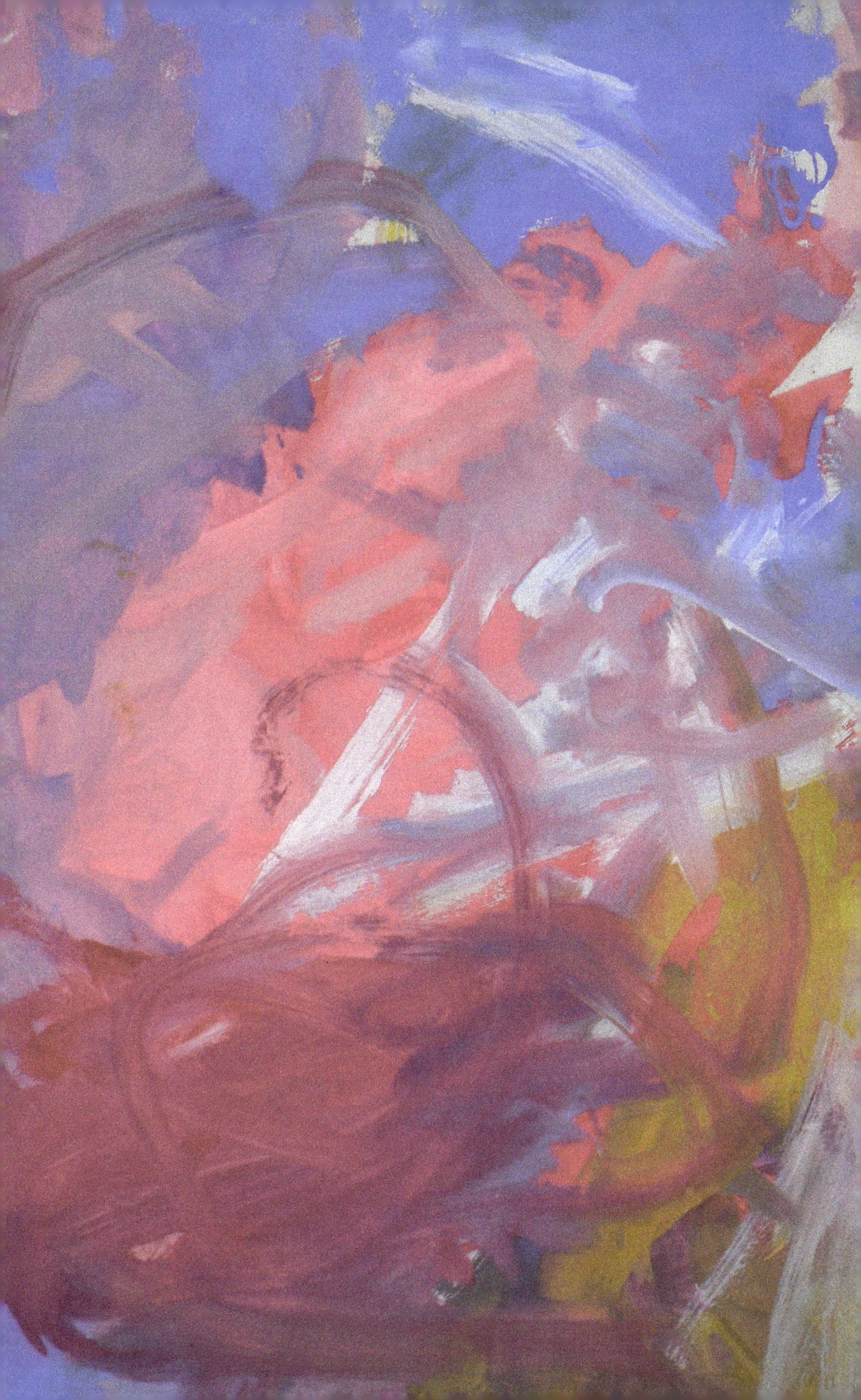

Cosmic Traveler

Who are you?
You appear, a nineteen-year-old, arriving out of the mist
Complaining about bad vibes
and a weird black dog that sent you on your way.
But you stop here.
My camp is always open, the fires lit
I wish you safe
Tonight, I will pray for you,
for your safe passage,
if only through here.

I am meant to serve as your oasis,
I hear the word to give you a place to stay.

You are riding the psychic wave of the land of the Anasazi
You don't yet know the extent of your knowledge
But you were led here, to a safe house
Your Guides led you here.

And if you are following your Guides,
You are on the Path.

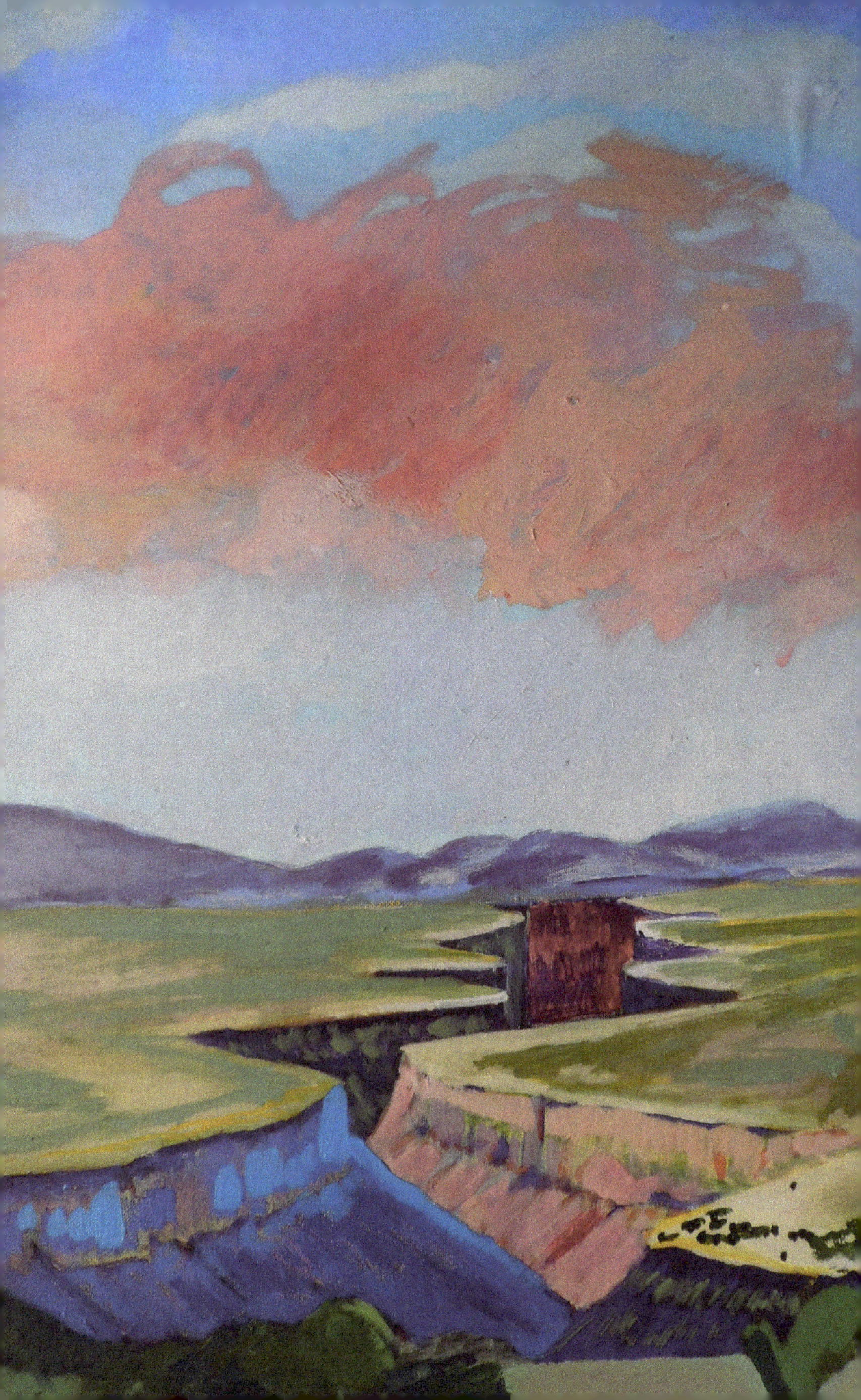

Companion Planets

Dear Sister
Side by side we walk through many lives
Our strength balancing each other
A touchstone of Love,
There is a special place in the Universe
For me to rest, when I need, on my journey
Thank you, dear Angel, for loving me.

Lunar Eclipse

A force of nature eclipses the Moon
Red Moon in the shadow of the Sun
shines copper in the sky,
Outside, muted chimes sound
Blocked, hidden, gone from view.

You move to eclipse my Moon, not illumine.
Celestial orbs move to darkness,
Music of the Spheres stops sounding
You are such a dark beauty.
However, movement continues
The eclipse always passes
I will still be here.
I stand in the dark with two feet on Mother Earth
Feeling the Wind as it fills me with the Chi of the Universe
And I listen to the rustling of the Trees
telling me of what is to come.